Adhesion and Bonding in Composites

Adhesion and Bonding in Composites

edited by

RYUTOKU YOSOMIYA
KIYOTAKE MORIMOTO
AKIO NAKAJIMA
YOSHITO IKADA
TOSHIO SUZUKI

CRC Press
Taylor & Francis Group
Boca Raton London New York

CRC Press is an imprint of the
Taylor & Francis Group, an **informa** business

First published 1990 by Marcel Dekker, Inc.

Published 2019 by CRC Press
Taylor & Francis Group
6000 Broken Sound Parkway NW, Suite 300
Boca Raton, FL 33487-2742

First issued in paperback 2019

© 1990 by Taylor & Francis Group, LLC
CRC Press is an imprint of Taylor & Francis Group, an Informa business

No claim to original U.S. Government works

ISBN 13: 978-0-367-45093-9 (pbk)
ISBN 13: 978-0-8247-8149-1 (hbk)

Visit the Taylor & Francis Web site at
http://www.taylorandfrancis.com

and the CRC Press Web site at
http://www.crcpress.com

Library of Congress Cataloging-in-Publication Data

Adhesion and bonding in composites / Ryutoku Yosomiya ... [et al.].
  p. cm.
  Includes bibliographical references.
  ISBN 0-8247-8149-X (alk. paper)
  1. Composite materials. I. Yosomiya, Ryutoku
TA418.9.C6A265 1989
620.1'18--dc20   89-36669
                      CIP

Preface

The remarkable progress of industrial technologies in recent years has made extremely high demands of materials of all types, including metals, polymers, and ceramics. In the aerospace field, for example, materials with lightness, high mechanical strength, and good heat resistance are required; whereas in the field of prostheses, such as artificial bones and blood vessels, good biological compatibility, including resistance to thrombus, as well as desirable mechanical characteristics are required.

With the increasing failure of single materials to respond to such high-level requirements, expectations for composites have risen. Composites are produced by combining two or more homogenous materials such that the physically and chemically different phases that are formed will make possible the high performance required. In achieving this high performance, the interaction between component materials in the interface, and the properties of each constituent, are of utmost importance. This book deals with the roles played by the component material interface in composites, with special emphasis on methods used to improve the adhesion and bonding between them.

Chapter 1 opens with a theoretical treatment of the interface in composites, based on previous studies. In Chapter 2 we describe the relationship between wetting properties and adhesion, primarily of polymer matrices. In many composites, sufficient wetting of the surface of a solid phase during a solid–liquid combining step is the basic condition for good adhesion (bonding). In composites, molecules and atoms of different materials get so close to each other as to form chemical bonds that contribute to produce a strong interaction. For this effect, control of the interfacial reaction is a requisite, and various treatments, such as coatings, surface modifications of reinforcing materials, or improvements of matrices, have been used.

Chapters 3 through 6 are devoted to descriptions of such surface modifications of matrices or fillers. Chapter 3 deals with modifications of matrix polymers through chemical and physical treatments, and Chapter 4 covers modifications through graft polymerization. In Chapter 5 we describe modifications of inorganic filters, and in Chapter 6 we discuss modifications of polymer matrices through polymer blending.

In Chapter 7 adhesion of plastics onto metals is dealt with, and in Chapter 8, bonding of ceramics with metals. Chapter 9 deals with modification of the interface in fiber-reinforced metal composites, and in Chapter 10 we discuss the interfacial effect in carbon-fiber-reinforced composites as related to methods of carbon fiber surface treatment. Chapter 11 discusses methods, with examples, in the interface analysis of composites, and Chapter 12 covers dynamic analysis and methods of measuring the interfacial strength of fiber-reinforced composites.

We believe that the book will be of value to scientists and engineers in the materials field and to engineers working with adhesives or composites, as well as people who are now students of composites.

The authors express their sincere thanks to numerous researchers whose excellent reports provided reference materials for this book. The authors are greatly indebted to Dr. Maurits Dekker, Chairman of the Board, Marcel Dekker, Inc., for his kind recommendation to publish this book, and are also deeply grateful to the editorial staff.

Ryutoku Yosomiya
Kiyotake Morimoto
Akio Nakajima
Yoshito Ikada
Toshio Suzuki

Contents

Contributors

RYUTOKU YOSOMIYA Department of Industrial Chemistry, Chiba Institute of Technology, Narashino-shi, Chiba, Japan

KIYOTAKE MORIMOTO Department of Composites, Tokyo Research Center, Nisshinbo Industries, Inc., Adachi-ku, Tokyo, Japan

AKIO NAKAJIMA Department of Applied Chemistry, Osaka Institute of Technology, Osaka-shi, Osaka, Japan

YOSHITO IKADA Director of the Research Center for Medical Polymers and Biomaterials, Kyoto University, Kyoto-shi, Kyoto, Japan

TOSHIO SUZUKI Director and General Manager of the R&D Division, Nisshinbo Industries, Inc., Chuo-ku, Tokyo, Japan

Adhesion and Bonding in Composites

1

Interfacial Characteristics of Composite Materials

1.1   INTRODUCTION

Composite materials are materials of composite structure comprising
two or more components that differ in physical and chemical properties which have been combined to provide specific characteristics for
particular uses. The boundaries between components are referred to
as solid interfaces.

Recently, it has become popular to consider the boundary not as
a contact interface without thickness but as an interphase with thicknesses on both components. That is, the concept involves the existence of a chemical and physical transient region or gradient in the
boundary. For example, a fiber-reinforced composite material includes
three phases: the surface of the fiber side, the interface between
the fiber and the matrix, and the interphase. These phases are referred to collectively as the interface.

The characteristics of the interface are dependent on the bonding
at the interface, the configuration, the structure around the interface, and the physical and chemical properties of the constituents.
As a result the interface has a strong influence on the property of
the composite material. In this chapter we describe interaction of
the interface and structural change in the interface.

1.2   COMPOSITE EFFECT AND INTERFACES

A method for the estimation of composite material performance from
the characteristics of fillers (reinforcing material: fiber, powder)
and matrices (polymer, metal, ceramics) and from the configuration
of the filler is generally called a law of mixture. In the most basic
form of a law of mixture, some characteristics of a composite material

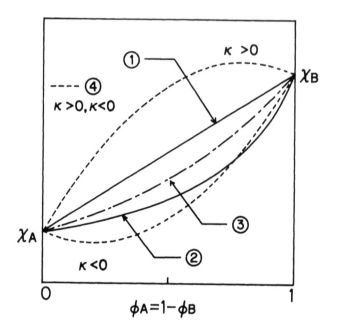

Figure 1.1   Relation between the properties of composites and various laws of mixture.   (Ref. 1)

are represented as a function of characteristics of constituent components and their volume fractions, as shown in Figure 1.1 [1]. For a composite material (characteristics: χ_c) that consists of component A (characteristics: χ_A, volume fraction: ϕ_A) and component B (characteristics: χ_B, volume fraction: ϕ_B), the basic formulas of a law of mixture are as follows:

$$\chi_c = \phi_A \chi_A + \phi_B \chi_B \qquad\qquad (1)$$

(parallel model, linear law of mixture, curve 1 in the figure) and

$$\frac{1}{\chi_c} = \frac{\phi_A}{\chi_A} + \frac{\phi_B}{\chi_B} \qquad\qquad (2)$$

(series model, curve 2 in the figure). The two curves exhibit theoretical upper and lower limits, respectively, based on a simple composite effect in general. A basic formula that generalizes (1) and (2) is

$$\chi_c^n = \phi_A \chi_A^n + \phi_B \chi_B^n \tag{3}$$

wherein $n(-1 \leqslant n \leqslant 1)$ represents a structural parameter which indicates the proportion of the combination mode; that is, the parallel mode is predominant when n is close to 1 and the series mode is predominant when n is close to -1. In equation (3), when n has a small absolute value,

$$\log \chi_c = \phi_A \log \chi_A + \phi_B \log \chi_B \tag{4}$$

This function is intermediate in behavior between the parallel model and the series model (curve 3 in the figure, referred to as a logarithmic law of mixture).

The law of mixture described above is valid for a simple composite system with well-known structure in which the rule of additivity holds (no interaction in the interface and no particular interfacial structure). However, it is natural to consider that in practice, any interaction will occur in the interface due to the contact between A and B. Then, considering the creation of interfacial phase C, different from components A and B, the following equation can be presented:

$$\chi_c = \phi_A \chi_A + \phi_B \chi_B + \kappa \phi_A \phi_B \tag{5}$$

Equation (5) represents a quadratic curve with a maximum (k > 0) or minimum (k < 0) depending on the sign of k (curve 4 in the figure, referred to as a quadratic law of mixture). The parameter k involves an interaction between components A and B and provides an expression of the interfacial effect. The equation suggests that the properties of the interfacial phase must be improved to obtain an excellent composite material.

On the other hand the properties of the interfacial phase are probably affected by the thickness of the phase and the intermolecular force between different molecules. To improve the properties of the interfacial phase, it is considered effective to increase the thickness of the interfacial phase and the intermolecular force between different molecules. However, a thick phase is not always formed at the interface. For example, a thick phase does not form in the case of polymer-metal differing from in the case of polymer-polymer [2]. Even when the interfacial phase is not thick, it is desirable to improve the interaction between different molecules, or to form primary bonding between molecules, to improve the properties of the interfacial phase.

Next is ion bonding and hydrogen bonding, and van der Waals bonding is lowest among interactions. However, some interaction always occurs, which is important in providing an interface, regardless of the type of material involved.

Furthermore, for interaction to be effective, other factors, such as a decrease in the distance between interfaces and an increase in the number of interactions, must be considered. These factors correspond to an improvement in the wettability between different constituent molecules of the interface. That is, thermodynamic wettability must be taken into account as the principal factor in the affinity between molecules. Wettability and adhesion are discussed in Chapter 2.

1.3 STRUCTURE AND FUNCTION OF INTERFACES

When a composite material (FRP) is formed, at least one fluid component in a form of solution or melt is mixed with other components and the mixture solidified. The structure formed at the interface with a contact medium is different to some degree from the internal structure. In general, bonding at the interface is described in terms of intermolecular force and surface free energy, but in practice the following factors are also important as determinant factors of interfacial bonding: (1) wettability, (2) chemical reaction, (3) adsorption and diffusion, (4) residual stress layer, (5) surface morphology, and (6) roughness effect. Factors that are important when combining inorganic materials and polymers (composite material) are discussed next.

1.3.1 Modification of Surface and Interfacial Interaction

In general, inorganic materials have a high surface free energy and organic materials a low surface energy. For favorable combination when the affinity is low, each material requires surface treatment. (In Chapters 3 through 6 the surface treatment of organic and inorganic materials and their effects are described.) Silane and esterification treatment are widely used as surface treatments for inorganic materials. For example, in Table 1.1 [3] are shown the critical surface free energies of glass, the surface of which is treated with various silane treating agents (silane coupling agent treatment). From the data in the table it is obvious that the high surface energy of glass is changed to a low critical surface free energy in a range near that of polymers. One reason for the application of such a treatment for reinforcing resins with glass fibers (FRP) is to decrease the critical surface free energy.

In addition, the adhesion strength depends to a considerable degree on the pH of a silane treatment solution. Silicon wafers were treated with γ-aminopropyl triethoxysilane solutions of various pH values and then bonded with polyimide resin. The influence of pH

Table 1.1 Critical Surface Tensions of Various Silane Coupling Agents on Borosilicate and Quartz Slides

Coupling agent and structure	Glass substrate	γ_c (dyn/cm)
3-(1,1-Dihydroperfluorooctoxy)propyltriethoxysilane, $CF_3(CF_2)_6CH_2O(CH_2)_3Si(OCH_2CH_3)_3$	Pyrex	14
γ-Perfluoroisopropoxypropyltrimethoxysilane, $(CF_3)_2CFO(CH_2)_3Si(OCH_3)_3$	Pyrex Silica	18 17
β-(p-Chlorophenyl)ethyltrimethoxysilane, p-$ClC_6H_4(CH_2)_2Si(OCH_3)_3$	Pyrex Silica	40–45[a] 44
γ-Chloropropyltrimethoxysilane, $Cl(CH_2)_3Si(OCH_3)_3$	Pyrex	43
Ethyltriethoxysilane, $CH_3CH_2Si(OCH_2CH_3)_3$	Silica	26–33[a]
Vinyltriethoxysilane, $CH_2{=}CHSi(OCH_2CH_3)_3$	Silica	30

[a]Depending on catalyst used.
Source: Ref. 3.

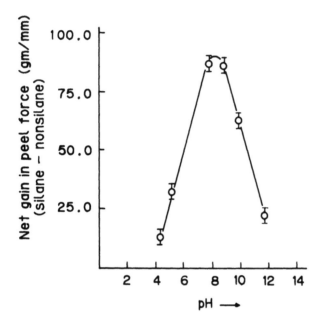

Figure 1.2 Peel force difference between silane- and non-silane-treated regions of a wafer plotted as a function of the pH of a γ-APS solution. The curve represents extrapolation of the data points. (Ref. 4.)

in treatment solutions is shown in Figure 1.2 [4]. Maximum adhesion strength is at a pH value near 9, and adhesion strength decreases when the pH is above or below this value.

It is assumed that polyimide reacts as follows when treated at a pH of 9:

(a)

With a pH below 9, aminosilane on the silicon surface forms intramolecular hydrogen bonding during treatment as shown in (b), and with a pH above 9, aminosilane forms bonds with hydroxyl groups on the silicon surface as shown in (c); consequently, functional groups

(b)   (c)

reactive with polyimide disappear and it is believed that this results
in a great decrease in adhesive strength.

Some say that organic material and polymer are chemically bonded
by functional groups on both ends of the silane treatment agent, but
Plueddeman considers the reaction between silanol groups and on the
surface of the glass and the silane treatment agent to be reversible.
In that case, the internal stress relaxation and hydrogen bonding
through a sliding mechanism greatly affect the retention of bonding
in the interface [5].

Recent investigations by Ishida et al. using an ingenious FT-IR
method have clarified the structure of silane-treated layers [6].
That is, silane-treated layers do not have a simple structure that
permits silane coupling agent molecules to chemically bond on the
glass surface in the form of monomolecular layers; rather, it forms
a more complex stratified structure. These multimolecular structure
changes are not the result of the chemical structure of the silane
treatment agent only, but also of the pH of the treatment solution,
the concentration, the solvent, and the temperature, in addition to
the surface structure of the reinforcing material (inorganic material).

Because the usual silane coupling agents will not react with matrices having no functional group, such as polyethylene and polypropylene, special treating agents having molecular structures that are
reactive with such matrices have been developed [7]. In Figure 1.3
the reaction mechanism of azidosilane are shown. Properties of polypropylene incorporated with three types of inorganic powder fillers
treated with azidosilane are shown in Table 1.2. In all cases the
tensile strength, bending strength, and heat deformation temperature
are improved.

Aside from surface modification of fillers (described in Chapters
5 and 10), chemical and grafting treatments other than coupling treatments have been used to activate the surface. Recently, oxidation
treatments such as wet HNO_3 oxidation, dry air heating oxidation,
and anodic oxidation of commercial carbon fibers (CF) have been used
by Fitzer in a comparative investigation of various viewpoints concerning adhesion between the fiber and the matrix thermosetting resin in
composite materials [8].

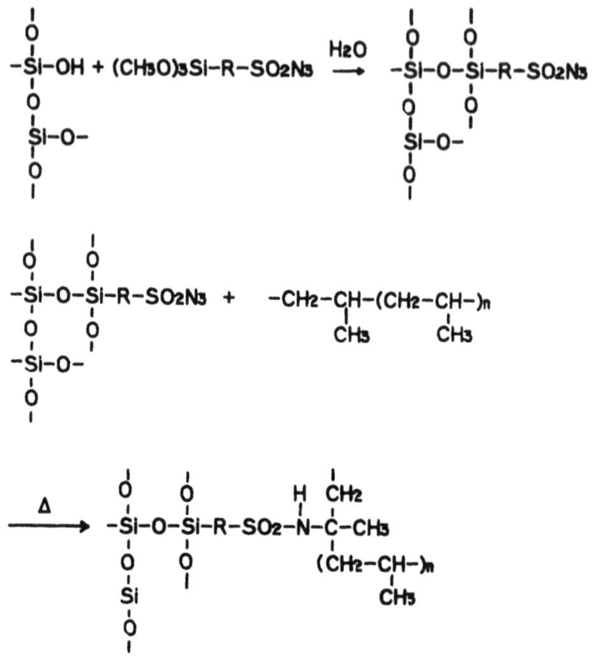

$$-\underset{\underset{Si-O-}{\overset{O}{|}}}{\overset{\overset{O}{|}}{Si}}-OH + (CH_3O)_3Si-R-SO_2N_3 \xrightarrow{H_2O} -\underset{\underset{Si-O-}{\overset{O}{|}}}{\overset{\overset{O}{|}}{Si}}-O-\underset{\overset{O}{|}}{\overset{\overset{O}{|}}{Si}}-R-SO_2N_3$$

$$-\underset{\underset{Si-O-}{\overset{O}{|}}}{\overset{\overset{O}{|}}{Si}}-O-\underset{\overset{O}{|}}{\overset{\overset{O}{|}}{Si}}-R-SO_2N_3 + -CH_2-\underset{\overset{|}{CH_3}}{CH}-(CH_2-\underset{\overset{|}{CH_3}}{CH}-)_n$$

$$\xrightarrow{\Delta} -\underset{\underset{Si}{\overset{O}{|}}}{\overset{\overset{O}{|}}{Si}}-O-\underset{\overset{O}{|}}{\overset{\overset{O}{|}}{Si}}-R-SO_2-\underset{\overset{|}{N}}{N}-\underset{\overset{|}{(CH_2-CH-)_n}}{C}-CH_3$$

Figure 1.3   Addition reaction of an azidosilane coupling agent to an inorganic filler.   (Ref. 7.)

The relation between the degree of oxidation, BET surface area, and interlaminar shear strength (ILSS) of composite materials is shown in Figure 1.4.   A maximum ILSS value is found for high-tenacity CF (Sigrafil HT); for high-modulus CF (Sigrafil HM) the ILSS value increases with increasing oxidation.   The behavior depends on the difference in surface structure of the CF.   Next, the effect of blocking of various functional groups resulting from oxidation treatment on the ILSS of composite material is determined. Acidic groups are blocked by diazomethane treatment, nonacidic hydroxyl groups by dimethyl sulfate treatment and carbonyl, and quinone groups by $NaHB_4$ and subsequent dimethyl sulfate or diazomethane treatment, and blocked CFs are combined with epoxy resin to form composite materials.   ILSS is shown in Figure 1.5.   For all functional groups involved in this experiment, some interaction between matrices is found.

Complete blocking of all functional groups generated results in ILSS values as low as that of the original untreated fiber.   Chemical interaction between the matrix and various functional groups resulting

Table 1.2   Comparative Physical Properties of Polypropylene[a] with 40 weight percent filler with and without aziodosilane coupling agent S3046[b]

Filler type	Unfilled	Mica[c]		Wollastonite[d]		Clay[e]	
Coupling agent	None	None	S3046	None	S3046	None	S3046
Tensile strength (psi), D638-72	4,300	4,370	7,190	3,960	4,980	3,260	4,330
Breaking elongation (%)	575	2.5	3.0	10	4.4	2.7	4.4
Tensile modulus (10^3 psi)	141	580	630	470	480	257	320
Flexural strength (psi), D790-71	5,800	7,230	12,800	7,700	9,500	6,300	8,270
Flexural modulus (10^3 psi)	180	780	1,030	620	700	350	460
Impact strength (ft-lb/in.) D256A-73 (notched izod)	0.49	0.49	0.44	0.61	0.48	<0.20	0.44
Heat deflection temperature at 264 psi (°F), D648-72	130	221	261	200	222	157	167
Density (g/cm^3)	0.92	1.225	1.238	1.245	1.260	1.224	1.242

Source: Ref. 7.
[a]Profax 6523 polypropylene, Hercules Inc.
[b]Azidosilane S3046, a development product of the Organics Department, Hercules Inc.
[c]Mica Alsibronz 12, 325 mesh, Franklin Mineral Corp.
[d]Wollastonite (calcium silicate), grade F-1, Interpace Corp.
[e]Clay-Suprex clay, 325 mesh, J. M. Huber Corp.

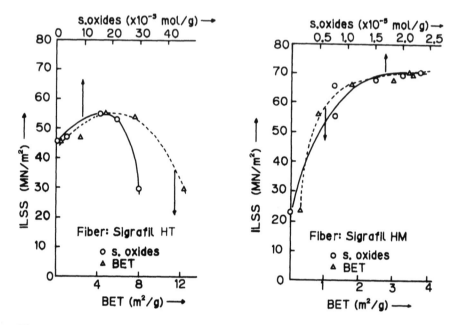

Figure 1.4 Correlation of interlaminar shear strength (ILSS) increase after oxidation with fiber BET surface area and amount of surface oxide. (Ref. 8.)

from oxidation treatment of CF is related strongly to improved adhesion between CF and the matrix. Possible chemical interactions such as these shown in Figure 1.6 are suggested.

Surface modification of polymer for adhesion is described fully in Chapters 3 and 4. We can summarize it here by saying that various functional groups are formed on the surface of a polymer to change its high-energy state. For example, flame treatment is a method of modifying the surface, in which a perfect combustion gas of mixed air or oxygen and natural gas containing methane and propane mixed homogeneously in a certain proportion is blown against the surface to be modified. When surface chemical structure of flame-treated low-density polyethylene (LDPE) was analyzed by XPS it was found that a considerable number of oxygen-containing groups were introduced on the surface [9].

In Table 1.3 the relation between treatment condition, introduced oxygen-containing group, and adhesion is shown. The adhesion increases with increased oxygen-containing groups. Ultraviolet (UV) radiation is also effective in surface modification. The surface chemical structure change of poly(ethylene terephthalate) (PET) by UV

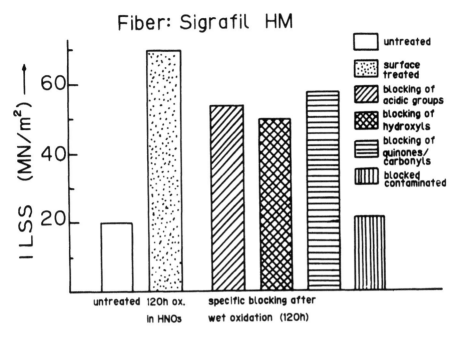

Figure 1.5   Contribution of nonacidic oxides to the strength ILSS of composites with high-modulus-carbon fibers.   (Ref. 8.)

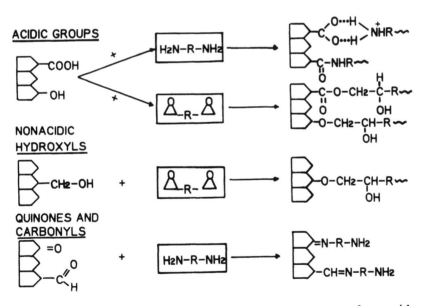

Figure 1.6   Possible relationships of carbon fiber surface oxides with epoxy and amine-containing chemicals.   (Ref. 8.)

Table 1.3 XPS and Joint Strength Data for Polyethylenes That Have Been Flame Treated

Polymer	Treatment time (s)	Natural gas flow (cm^3/s)	Airflow (cm^3/s)	O:C (at. %)	N:C (at. %)	Lap shear[a] strength (MN/m^2)	Standard deviation of sample
Alkathene 47[b]							
	0	0	0	0.25	0	0.55	0.07
	1.2	37	150	16.9	0.94	6.6	0.6
	4.8	37	150	31.0	3.2	7.2	0.7
	1.2	74	317	15.3	2.2	6.8	0.7
	1.2	18.5	75	6.8	0	5.1	0.6
Alkathene 11[c]							
	0	0	0	<0.25	0	0.36	0.04
	1.2	37	150	20.5	1.5	5.6	0.5
	4.8	37	150	33.4	3.2	7.2	0.4
	1.2	74	317	13.7	2.5	6.4	0.3
	1.2	18.5	75	5.1	0	5.7	0.4

[a]With treated polymers the failure was always a mixture of apparent interfacial and material.
[b]Contains no additives.
[c]Contains 0.02% 2,6-di-tert-butyl-p-Cresol.
Source: Ref. 9.

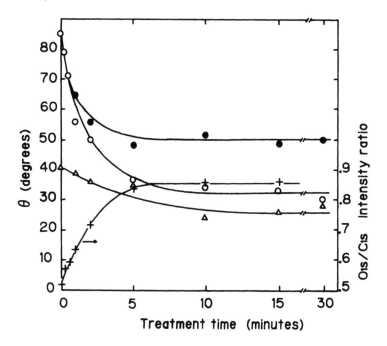

Figure 1.7   Contact angle of water (o) and methylene iodine (Δ) on PET as a function of treatment time. Water contact angle (•) on specimens washed before measurement. ESCA O_{1S}/C_{1S} intensity ratio as a function of photooxidation time (+). (Ref. 10.)

radiation in an oxygen atmosphere has been analyzed by ESCA, and the result in terms of wettability is shown in Figure 1.7 [10]. O_{1s}/C_{1s} increases in proportion to treatment time, and in responding to the increase, θ decreases and the wettability is improved.

As shown in Figure 1.8, θ increases over several days after treatment by being allowed to stand, and in responding its surface energy decreases. The factor that dominates the decrease is the polar component; the dispersion component does not change essentially. As described in Chapter 2, it is believed that polar groups formed on the surface penetrate the resins in such a manner as to cause such a phenomenon. Recent progress in the development of analytical equipment has facilitated the investigation of details of surface structure and surface phenomena.

1.3.2   Formation of Higher-Order Structure at the Interface (Structural Change in the Matrix at the Interface)

When a crystalline polymer is solidified from melt at a suitable temperature and in contact with a suitable medium, lamella crystals several

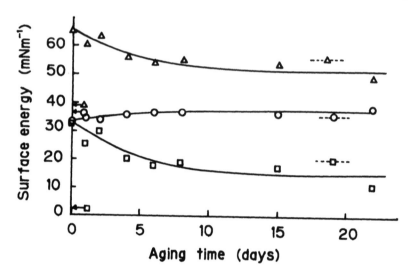

Figure 1.8   Surface free energy (△), surface free energy due to
polar forces (□), and surface free energy due to dispersion forces
(○) for a sample given 30-min treatment as a function of aging time.
Values for washed samples are indicated for each energy term by a
dashed line through the appropriate symbol.   The arrows along the
energy axis indicate the energy terms for untreated PET.   (Ref. 10.)

tens of micrometers thick of uniaxial orientation are formed at the
interface.   This structure, which differs from the internal structure, is usually called the transcrystalline region [11].   For example,
the relation between wettability and the crystallization of polyethylene when melted polyethylene is solidified on various solid surfaces
is shown in Table 1.4.   The γ_c value, which is an indicator of
wetting characteristics, and the wettability increase with increasing
crystallinity.   Such surface crystallization depends on the polarity
of molding material; when molding is carried out on a solid (material)
with high surface energy, the crystallization is accelerated and the
wettability of the surface is improved.
    In the process of injection molding of polypropylene, a crystal-
line polymer, the dependence of peel adhesion to metal on both mold
material and cooling rate is shown in Table 1.5 [12].   When polar
materials such as aluminum are used and the cooling rate is slow,
peel adhesion is high.   It is seen by electron microscopic observation of cross sections that when molding is carried out using a mold
of a polar material such as aluminum, spherulite structure grows
from the surface.   When less polar materials such as PET are used

Table 1.4   Wettability of Melt-Crystallized Polyethylene Film Formed by Nucleation at Solid-Liquid and Liquid-Vapor Interfaces[a]

	θ (deg)	$\cos\theta$	ρ_s (g/cm^3)	V_{sp} (cm^3/g)	γ_c (dyn/cm)	Crystallinity (%)
Polytetra-fluoroethylene	81	0.1564	0.855	1.1695	36.2	0
Mylar	81	0.1564	0.855	1.1695	36.2	0
Cu	80	0.1737	0.862	1.1601	37.4	5.1
Ni	68	0.3746	0.933	1.0718	51.3	53.3
Sn	66	0.4067	0.944	1.0593	53.8	60.1
Al	65	0.4226	0.949	1.0537	54.9	63.2
Glass	65	0.4226	0.949	1.0537	54.9	63.2
Cr	64	0.4384	0.954	1.0482	56.1	66.2
Au	53	0.6018	1.007	0.9930	69.6	93.6
NaCl	57	0.5446	0.989	1.0111	64.8	86.4
KBr	57	0.5446	0.989	1.0111	64.8	86.4
KCl	57	0.5446	0.989	1.0111	64.8	86.4
CaF$_2$	55	0.5736	0.998	1.0020	67.3	91.4

[a]Polyethylene was in contact with surfaces for 30 min at 200°C. All measurements were performed at 20° using glycerol as the wetting liquid.
Source: Ref. 11b.

Table 1.5   Peel Adhesion for Polypropylene[a] as a Function of
Compression Molding Variables

Molding variables[b]		Peel adhesion[c] (lb/in.)	
Cooling rate	Mold surface	As molded[d]	Postannealed[e]
Slow	Aluminum	16	17–20
Normal	Aluminum	12	17–20
Fast	Aluminum	0–2	14–17
Normal	Mylar	2–3	7–8

Source: Ref. 12.

[a] A homopolymer (Hercules Profax 6523) was used.

[b] Samples were compression molded at 210°C for 15 min and then
colled to room temperature under pressure overnight (slow), to 65°C
under pressure in 20 min (normal), or to approximately 10°C without pressure in 15 s by immersion in ice water (fast).

[c] All samples were plated under the preferred conditions described in
the experimental section. Each adhesion value is an average of at
least five separate peel tests.

[d] Samples were tested 1 to 2 days after plating.

[e] Samples were heated for 1 h at 80°C (1 to 2 days after plating)
and allowed to stand overnight at room temperature before testing.

the transition at the surface layer is from amorphous to transcrystalline, and internal typical spherulite structures are formed.

   Some investigations have addressed the question of improvement
in FRP properties by such topotactic reactions [13]. For example,
graphite fibers are ideal materials to use as the crystallization nucleation agent because of their relatively higher surface energy. Crystals of nylon 6,6, nylon 6, or polyethylene grow from the fiber surface to form a strong adhesion with graphite fibers. A comparison
of the properties of polycarbonate reinforced with graphite fibers
and with glass fibers is given in Table 1.6.

   It is obvious that the tensile strength and modulus of fiber-
reinforced polycarbonates are greatly improved by heat treatment,
but not of the polycarbonate itself; the effect of heat treatment is
more remarkable for reinforcement with graphite fibers. Elevation

Table 1.6   Effect of Heat Treatment on the Mechanical Properties of Polycarbonate Composite

	Heat treatment		
	190(°C)	275(°C)	275/245[a] (°C)
Glass fiber reinforced			
Tensile strength (psi × 10³)			
Non-fiber reinforced	9.0	9.1	8.9
Fiber reinforced	7.6	11.1	11.2
Elongation (%)			
Non-fiber reinforced	5.1	5.3	4.9
Fiber reinforced	1.3	2.0	2.3
Young's modulus (psi × 10⁵)			
Non-fiber reinforced	3.25	3.30	3.11
Fiber reinforced	8.19	8.20	9.28
Glass transition temperature (°C)			
Non-fiber reinvorced	145–150	145–150	145–150
Fiber reinforced	147–148	143–144	145–146
Graphite fiber reinforced			
Tensile strength (psi × 10³)			
Non-fiber reinforced	8.3	8.6	8.2
Fiber Reinforced	8.6	13.7	16.3
Elongation (%)			
Non-fiber reinforced	5.1	6.2	2.8
Fiber reinforced	0.8	0.8	0.8
Young's modulus (psi × 10⁵)			
Non-fiber reinforced	3.1	3.4	3.3
Fiber reinforced	19.3	29.0	37.8
Glass transition temperature (°C)			
Non-fiber reinforced	150	150	150
Fiber reinforced	164	162	164

[a]Heat treated at 275°C and annealed at 245°C.
Source: Ref. 13a.

of the glass transition temperature and the increase in modulus are presumed to be attributable to crystallization on the fiber interface.

As mentioned above, heterogeneous physical structure is found in crystalline polymers; similarly, heterogeneous chemical structure is found in multicomponent polymers. In particular, in multiphase polymers such as polymer blends, block polymers, and grafted polymers, many surface concentration phenomena have been found through progress in surface analysis techniques. Recently, such phenomena have been found in random copolymers and homopolymers and have attracted both academic and industrial interest.

It is clear that the heterogeneous chemical structure between bulk and surface depends on (1) the difference in surface tension of polymer components and (2) molding conditions, especially the mold material (discussed in Chapter 6). An example of homopolymers is described here.

The contact angle with water of both surfaces of poly(butyl methacrylate) (PBMA) films (the film being formed on various substrates by a solvent casting method), was measured, and the results obtained are shown in Table 1.7 [14]. Air-side surfaces (the "top" surface) exhibit almost the same θ value as shown in the table, regardless of different substrates, which suggests the existence of the same chemical structure. On the other hand, substrate-side surfaces (the "bottom" surface) exhibit different θ values depending strongly on the substrate as shown in the table; that is, the θ value decreases in the order polytetrafluoraethylene (PTFE; Teflon) > glass > water, and that means increasing wettability.

These facts indicate that the surface structure is formed corresponding to the chemical structure of a substrate used for film preparation, and probably that carbonyl groups of PBMA molecules are oriented on the adjacent surface layer of substrate to form a surface structure with a polar substrate such as water.

To form such an orientation of polar groups of PBMA molecules on the surface of a polar substrate, free movement of polymer chains is important. Then a film must be formed from a polymer solution or polymer melt with a temperature higher than its T_g to accelerate molecular motion. For instance, PBMA film cast on glass exhibits a θ_A value of 65°, as shown in Table 1.7, corresponding to that of polarized structure. But when the film was heated at 50°C (higher than the T_g of PBMA) on the surface of Teflon, the wettability of the surface changed to 83°, as shown in Table 1.8 by the bottom θ_A value suggesting that a change to nonpolar structure can occur on a Teflon surface.

On the other hand, when the film is put in contact with a Teflon surface at 8°C, the film exhibits a θ_A value of 66°, suggesting retention of the surface structure formed on the glass surface as it is. Treatment on the glass surface of PBMA film with a hydrophobic surface structure prepared on a Teflon surface gives the same results.

Table 1.7 Contact Angles for Water on Poly(butyl methacrylate) (PBMA) Films

Substrate	Top surface		Bottom surface	
	θ_A[a] (deg)	θ_R[b] (deg)	θ_A[a] (deg)	θ_R[b] (deg)
PTFE	85.5	70.0	85.0	68.0
Glass	85.5	69.5	65.0	59.0
Water	83.5	71.5	63.5	59.5

[a] θ_A, Maximum advancing angles.
[b] θ_R, Minimum receding angles.
Source: Ref. 14.

Table 1.8 Contact Angles[a] for Water on Poly(butyl methacrylate) (PBMA) Films After Heating or Cooling on Different Substrates for 5 h

System	Top surface		Bottom surface	
	θ_A (deg)	θ_R (deg)	θ_A (deg)	θ_R (deg)
Film cast on glass; then heated at 50°C on PTFE	84	70	83	69
Film cast on glass; then cooled at 8°C on PTFE	85	70	66	60
Film cast on PTFE; then heated at 50°C on glass	85.5	71	72	63
Film cast on PTFE; then cooled at 8°C on glass	84	70	83	65

[a] All angles measured at room temperature
Source: Ref. 14.

These facts suggest the acceleration of molecular motion by heating at a temperature higher than the T_g of the polymer and consequent rearrangement of the polar components of the polymer, corresponding to the polarity of the substrate, to construct a new surface structure. Since the chemical structure of bulk matrix polymer can be different from that of the interfacial (surface) matrix polymer, depending on the polarity of fillers in a composite material, it is natural to assume that the polarity affects the property of the composite.

Recently, many analyses of the surface and interface have been carried out using new analytical techniques such as FT-IR and XPS. For example, the fiber-matrix interface of poly(methyl methacrylate) (PMMA) reinforced with Kevlar fibers was analyzed to clarify the interaction between carbonyl groups of PMMA and amide groups of Kevlar and to describe the interfacial structure. As a result it has become obvious that the interfacial structure depends greatly on the tacticity of the matrix and that many carbonyl groups of PMMA exist on the fiber interface in the case of atactic and syndiotactic PMMA [15,16].

REFERENCES

1a. T. Hata, *Composite Materials*, University of Tokyo, Shuppankai, 1975, p. 17.

1b. I. Kimpara, *Composite Materials*, High Polymers, Jpn, *31*, 1081 (1982).

2a. J. Letz, Diffuse Interphase Layer in Microheterogeneous Polymer Mixtures, *J. Polym. Sci. A-2*, 7, 1987 (1969).

2b. J. Letz, Thickness of the Interphase Layer Determined from Volume Change Produced by Mutual Diffusion in Binary Microheterogeneous Polymer Mixtures, *J. Polym. Sci. A-2*, *8*, 1415 (1970).

3. W. A. Zisman, Surface Chemistry of Plastics Reinforced by Strong Fibers, *Ind. Eng. Chem. Prod. Res. Dev.*, *8*, 98 (1969).

4. D. Suryanarayana and K. L. Mittal, Effect of pH of Silane Solution on the Adhesion of Polyimide to a Silica Substrate, *J. Appl. Polym. Sci.*, 29, 2039 (1984).

5. E. P. Plueddemann, Mechanism of Adhesion of Coating Through Reactive Silanes, *J. Paint Technol.*, 42, 600 (1970).

6. H. Ishida and G. Kumareds, *Molecular Characterization of Composite Interfaces*, Plenum Press, New York, 1985, p. 25.

7. G. A. McFarren et al., Azidosilane Polymer-Filler Coupling Agent, *Polym. Eng. Sci.*, *17*, 46 (1977).

8. E. Fitzer and R. Weiss, Effect of Surface Treatment and Sizing of C-Fiber on the Mechanical Properties of CRF Thermosetting and Thermoplastic Polymer, *Carbon*, *25*, 455 (1987).

9. D. Briggs, D. M. Brewis, and M. B. Konieczko, X-Ray Photoelectron Spectroscopy Studies of Polymer Surfaces, *J. Mater. Sci.*, *14*, 1344 (1979).

10. J. Peeling, G. Courval, and M. S. Jazzar, ESCA and Contact-Angle Studies of the Surface Modification of Polyethylene-Terephthalate Film, Photooxidation and Aging, *J. Polym. Sci. Polym. Chem. Ed.*, *22*, 419 (1984).

11a. K. Hara and H. Schonhorn, Effect on Wettability of FEP, Teflon Surface Morphology, *J. Adhes.*, *2*, 100 (1970).

11b. H. Schonhorn, Effect of Substrate on Morphology and Wettability, *Macromolecules*, *1*, 145 (1968).

11c. H. Schonhorn, Heterogeneous Nucleation of Polymer Melt on Surface, *Polym. Lett.*, *5*, 919 (1967).

12. D. R. Fitchmun, S. Newmann, and R. Wiggle, Electroplating on Crystalline Polypropylene, *J. Appl. Polym. Sci.*, *14*, 2441 (1970).

13a. F. S. Cheng, J. L. Kardos, and T. L. Tolbert, One Way to Strengthen Graphite/Polycarbonate Composite, *SPEJ*, *26*, 62 (1970).

13b. F. Tuinstra and E. Baer, Epitaxial Crystallization of Polyethylene on Graphite, *J. Polym. Sci. Lett.*, *B-8*, 861 (1970).

14. Z. Haq and J. Mirgins, The Contact Angle of Water on Polymer Films, *Polym. Commun.*, *25*, 269 (1984).

15. M. Kodama and K. Kuramoto, XPS Study of Boundary Phase Structure Between Stereoregular Polymethylmethacrylate and Polyamide Substrate, *Polym. J.*, *20*, 515 (1988).

16. M. Kodama and I. Karino, Effects of Polar Groups of Polymer Matrix on Reinforcement Matrix Interaction in Kevlar Fiber Reinforced Composites, *J. Appl. Polym. Sci.*, *32*, 5057 (1986).

2
Wettability and Adhesion

2.1 INTRODUCTION

A composite material is a combination of three or more materials and
its strength depends on the characteristics of the interface. In this
chapter we review the relationship between the wetting phenomenon
at the interface and the interfacial strength in terms of the interface chemistry.

The interfacial strength (adhesive strength) is determined basic-
ally by the cohesion force of the matrix itself and the interaction between the matrix and the filler surface. In general, the adhesion
phenomenon should be treated microscopically as an interfacial chemical
phenomenon. This idea is based on the following three factors [1]:

1. Mechanical binding, including anchor effect. For this purpose, it is necessary to increase the wet area by increasing
   the contact area.
2. Binding by physical interaction. Here, van der Waals forces
   are considered to be the principal intermolecular force. In
   other words, there is a permanent polar effect, an induced
   polar effect, and a dispersion effect such as the van der
   Waals force. Among these, the permanent polar effect is a
   dipole-dipole interaction and exists between polar molecules.
   The induced polar effect is an interaction between the dipole
   induced in the second molecule by the electric field of the
   polar molecule and the dipole of the polar molecule, and exists
   as the interaction between polar and nonpolar molecules. The
   dispersion effect is an interaction between nonpolar molecules.
3. Binding by chemical interaction, where the hydrogen bond,
   convalent bond, and others are considered.

In this chapter the polymer will generally be considered as the
matrix.

2.2 WETTABILITY AND ADHESION FORCE

When phases 1 and 2 contact an interface free energy (γ_{12}), the following relation exists among the energy required to separate them, W_a; the surface energy of phase 1, γ_1; the surface energy of phase 2, γ_2; and the interfacial free energy, γ_{12}:

$$W_a = \gamma_1 + \gamma_2 - \gamma_{12} \tag{1}$$

Since the concept of surface free energy is based on reversible equilibrium theory, W_a is reversible and is said to be the amount of work of thermodynamic adhesion. On applying W_a to the adhesion system, the following assumptions are necessary: (1) ideal separation at the interface and (2) no other energy changes (i.e., isothermal conditions).

When phase 1 is solid (S) and phase 2 is liquid (L), Young's equation,

$$\gamma_S = \gamma_{SL} + \gamma_L \cos \theta \tag{2}$$

can be used to give

$$W_a = \gamma_L (1 + \cos \theta) \tag{3}$$

That is, W_a between the solid and the liquid can be calculated if the surface tension of the liquid and the contact angle (θ) are known.

Although in a real adhesion system, the evaluation of W_a between solid and solid and the value of the surface free energy of the solid are required, there is no way to measure them directly such as in the case of a liquid. Therefore, an evaluation method based on the W_a value of the solid-liquid is employed. For example, Good and Girifalco [2] assumed the interface free energy as to be

$$\gamma_{12} = \gamma_1 + \gamma_2 - 2\phi(\gamma_1\gamma_2)^{1/2} \tag{4}$$

and obtained the surface free energy. Here ϕ is an interaction parameter called the work function of adhesion.

Comparing equation (1) and (4), the equation

$$W_a = 2\phi(\gamma_1\gamma_2)^{1/2} \tag{5}$$

is derived. Fowkes [3] assumed that the interfacial interaction force consists only of the dispersion force (d) and derived the following equation:

$$\gamma_{12} = \gamma_1 + \gamma_2 - 2(\gamma_1^d \gamma_2^d)^{1/2} \tag{6}$$

Extended Fowkes equations, which consider a polar component (γ^p) and a hydrogen bond component (γ^H), were proposed by Owens [4], Kaelble [5], Wu [6], and Kitazaki and Hata [7]:

$$\gamma_{12} = \gamma_1 + \gamma_2 - 2(\gamma_1^d \gamma_2^d)^{1/2} - 2(\gamma_1^p \gamma_2^p)^{1/2} \tag{7}$$

$$\gamma_{12} = \gamma_1 + \gamma_2 - 2(\gamma_1^d \gamma_2^d)^{1/2} - 2(\gamma_1^p \gamma_2^p)^{1/2} - 2(\gamma_1^H \gamma_2^H)^{1/2} \tag{8}$$

$$\gamma_{12} = \gamma_1 + \gamma_2 - \frac{4\gamma_1^d \gamma_2^d}{\gamma_1^d + \gamma_2^d} - \frac{4\gamma_1^p \gamma_2^p}{\gamma_1^p + \gamma_2^p} \tag{9}$$

From these equations, 1 and 2 are read as S (solid) and L (liquid), respectively, and

$$W_a = \gamma_S + \gamma_L - \gamma_{SL} = \gamma_L(1 + \cos \theta)$$

is introduced into the respective equation:

$$W_a = \gamma_L(1 + \cos \theta) = 2(\gamma_S^d \gamma_L^d)^{1/2} + 2(\gamma_S^p \gamma_L^p)^{1/2} \tag{10}$$

$$W_a = \gamma_L(1 + \cos \theta) = 2(\gamma_S^d \gamma_L^d)^{1/2} + 2(\gamma_S^p \gamma_L^p)^{1/2} + 2(\gamma_S^H \gamma_L^H)^{1/2} \tag{11}$$

$$W_a = \gamma_L(1 + \cos \theta) = \frac{4\gamma_S^d \gamma_L^d}{\gamma_S^d + \gamma_L^d} + \frac{4\gamma_S^p \gamma_L^p}{\gamma_S^p + \gamma_L^p} \tag{12}$$

When one measures the contact angel (θ) using a liquid having known γ_L^d, γ_L^p, and γ_L^H, equations (10) through (12) become combined equations that permit calculation of γ_S.

Table 2.1 shows the surface free energy of various polymers obtained in this way [7b and 8]. W_a is a thermodynamic index of adhesiveness.

Table 2.1 γ_s^a, γ_s^b, γ_s^c, and γ_s of Polymer Solids (dyn/cm, at 20°C)

Polymer	γ_s^a	γ_s^b	γ_s^c	γ_s	$\gamma_c(max)*,†$	$\gamma_c‡$
Hexafluoropropylene	14.9	0	0	14.9	–	16.2
Polytetrafluoroethylene	19.4	2.1	0	21.5	21.5 (B)	18.5
Polytrifluoroethylene	22.1	7.8	1.3	31.2	29.0 (C)	22
Poly(vinylidene fluoride)	27.6	9.1	3.5	40.2	40.0 (C)	25
Poly(vinyl fluoride)	42.3	0.2	1.0	43.5	44.2 (C)	28
Polyethylene	35.6	0	0	35.6	38.3 (B)	31
Poly(vinylidene chloride)	43.0	1.9	0.9	45.8	44.0 (B)	40
Poly(vinyl chloride)	43.7	0.1	0.2	44.0	43.9 (B)	39
Poly(methyl methacrylate)	42.4	0	0.8	43.2	43.2 (B)	39

	γ_S^a	γ_S^b	γ_S^c	γ_S	$\gamma_c(\max)$[*]	γ_c[‡]
Polyacrylamide	26.5	15.1	10.7	52.3	—	—
Poly(vinyl alcohol)	36.5	3.3	—	—	—	37
Polystyrene	33.8	5.8	0	40.6	43.0 (B)	33
Poly(ethylene terephthalate)	42.7	0.6	0.5	43.8	43.4 (B)(C)	43
Nylon	42.0	1.4	3.1	46.5	46.0 (C)	46
Hexatriacontane	20.6	0	0	20.6	20.6 (A)	20–22
Paraffin	24.4	0	0	24.4	25.7 (B)	23
Polypropylene	29.8	0	0	29.8	29.8 (B)	29
Polyoxymethylene	42.5	0.9	1.2	44.6	46.5 (C)	—
Poly(γ-methyl L-glutamate)						
α-sheet	38.8	1.0	8.2	48.0	50 (C)	40–50
β-sheet	33.3	2.4	2.1	37.8	37 (B)(C)	37

Source: Ref. 7b and 8.

[*]$\gamma_c(\max)$, maximum value of γ_c among the values obtained by the different liquid series A, B, and C.

[†]Letter in parentheses indicates type of liquid; $\gamma_S = \gamma_S^a + \gamma_S^b + \gamma_S^c$.

[‡]γ_c, critical surface tension reported in the literature.

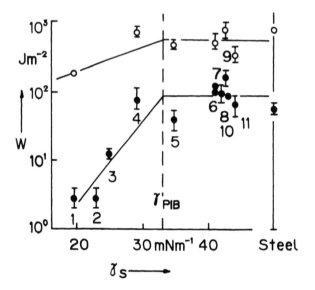

Figure 2.1   Adhesive failure energy W for two different contact periods, plotted against the surface tension of the polymers used as adherents.   T = 23°C; ●, t = 1, 5 × 10^{-2} s; ○, t = 1.0 × 10^2 s.   1, polytetrafluoroethylene; 2, polysiloxane; 3, poly(vinylidene fluoride); 4, polypropylene; 5, polyethylene; 6, poly(vinyl chloride); 7, poly-(methyl methacrylate); 8, polystyrene, 9, poly(ethylene terephthalate); 10, 6-polyamide; 11, 6,6-polyamide.   (Ref. 9.)

Therefore, the larger W_a, the better the adhesiveness. However, W_a estimated from this table is smaller than 100 dyn/cm and much smaller than the adhesion or cohesion break energy, 10^5 to 10^7 dyn/cm, which was measured mechanically.

    To examine the effect of wettability on adhesion strength, Zosel et al. [9] measured the adhesion strength of polyisobutylene (PIB) on various polymers and obtained the result shown in Figure 2.1. This figure shows the strength measured at contact times of 1.5 × 10^{-2} and 1 × 10^2 s. In both cases, the identical adhesion strength (w) is shown when the surface free energy, γ_S, of the adhered material is larger than the γ_S value of PIB. On the other hand, when the γ_S value of the adhered material is lower than the γ_S value of PIB (i.e., PIB does not wet the material enough), the adhesion strength increases when the γ_S value of the material approaches that of PIB.

    Recently, it has been reported for hot melt adhesion that good wettability does not necessarily contribute to adhesion strength.

In the hot melt adhesion of a polymer-polymer or polymer-metal system, Imachi [10] examined contact angle, peeling strength, hot melt temperature, and so on, and obtained the result that good wettability does not always bring high peeling strength. This may suggest that wettability is only one of the conditions of adhesion.

To examine the relationship between each component of the surface tension and the adhesion strength, Nakamae et al. deposited metallic cobalt vapor on various polymer films and measured the adhesion strength between a thin film of metallic cobalt and a polymer film [11]. The surface characteristics of the polymer film employed in this experiment are shown in table 2.2. The comparison shows that the dispersion component of various polymers does not change very much, but the polar component changes significantly. Such a balance between the dispersion and the polar component and the polar component itself greatly affects interaction between the polymer and the other adhered material.

Figure 2.2 shows the adhesion strength and the component of surface tension. A linear relation is observed between the adhesion strength and the polar component, but the dispersion component does not show such a relationship. Furthermore, the surface of the poly-(ethylene terephthalate) (PET) film was treated with aqueous NaOH solution. As shown in Figure 2.3, a relation between adhesion strength and each component of the surface tension was obtained using this treated film. In this case, also, a linear relation was observed between adhesion strength and the polar component.

From the results of surface analysis by FT-IR, the adsorption intensities of C=O and OH increased proportionally to treatment time. It is estimated from this that the strong polar interaction between highly polar metallic cobalt and polar substituents such as —OH, —COOH, or CO existing on the PET film surface contributes greatly to the increase in adhesion strength.

In the case of adhesion between inorganic material and polymer or between metallic material and polymer, various surface treatments are generally undertaken to get their surface free energy values close together. In this way, the mutual affinity is increased and the atmospheric strength, especially the waterproof strength, is increased.

For example, when various surface treatments are undertaken for aluminum, γ_S^d and γ_S^p change as shown in Table 2.3 [12]. The adhesion strength with epoxy phenol resin is superior for aluminum treated by electrolytic oxidation. However, the treated aluminum has a high adhesion strength to water, and the waterproof strength is inferior. On the other hand, aluminum treated with phosphoric acid is superior in waterproof strength. In this system, therefore, the γ_S^d value of aluminum should be large in order to raise the

Table 2.2 Characteristics of Various Polymer Films Used

Polymer	Contact angle							Surface free energy (dyn/cm)		
	Water			Ethylene glycol			Methylene iodine			
	θ_a	θ_r	θ	θ_a	θ_r	θ	θ	γ_s	γ_s^d	γ_s^p
Poly(ethylene terephthalate) PET	76	53	65	53	29	42	34	44	33	11
NaOH–treated poly(ethylene terephthalate)										
PET-H1	70	38	56	53	28	42	33	47	30	17
PET-H2	64	36	51	48	13	35	33	50	30	20
PET-H3	64	22	47	44	5	31	32	52	29	23
PET-H4	64	20	46	46	4	32	32	52	29	23

PET-H5	63	15	45	43	5	31	31	53	29	24
PET-H6	59	0	40	39	0	27	29	56	29	27
PET-H7	53	0	37	30	0	21	28	58	29	29
PET-H8	54	0	37	31	0	22	28	58	29	29
Polytetrafluoroethylene PTFE	109	79	98	86	70	78	70	22	20	2
Polypropylene PP	106	77	94	74	63	68	52	32	31	1
Polyethylene PE	101	76	88	70	49	60	40	36	34	2
Polystyrene PS	84	69	77	58	48	53	46	37	30	7
Vinyl chloride/vinyl acetate/vinyl alcohol copolymer P(VC-VAC-VA)	73	59	65	58	29	45	31	44	33	11
Poly(vinyl butyral) PVB	75	43	60	54	26	42	42	44	28	16
Poly(vinyl alcohol) PVA	53	27	42	29	0	18	42	55	26	29

Source: Ref. 11.

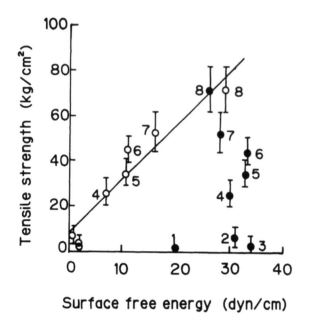

Figure 2.2 Relation between the tensile strength and the component of surface free energy of various substrates. •, Dispersion component; ○, polar component. 1, PTFE; 2, PP; 3, PE; 4, PS; 5, PET; 6, P(VC—VAC—VA); 7, PVB; 8, PVA. (Ref. 11.)

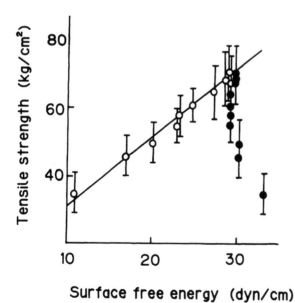

Figure 2.3 Relation between the tensile strength and the component of surface free energy of NaOH-treated PET films. •, Dispersion component; ○, polar component. (Ref. 11.)

Table 2.3 Surface Characteristics of Aluminum (mJ/m^2) and Reversible Energy of Adhesion of Aluminum to Epoxyphenolic Coating and Water (mJ/m^2)

Surface treatment	γ_S^D	I_{SW}^P	γ_S^P	W_0 Al/coating	W_0 Al/water
Hexane extraction	42	38.7	7.5	103	99
DMF extraction	135	62.5	19	182	170
Phosphatization	150	18	1.5	169	131
Anodization	125	95	44	176	198
Sealed anodization	41	55.5	15	101	115

Source: Ref. 12.

adhesion strength under dry conditions, while treatment to decrease the γ_S^P value is desirable to raise the waterproof strength.

These phenomena are considered in the light of adhesion work. When A—B forms the interface, the adhesion work in air and in liquid is defined as W_{AB} and W_{AB}^L, respectively. The results for aluminum oxide—polyethylene and stearic acid-treated aluminum oxide--polyethylene systems are shown in Table 2.4 [13]. For the case of no treatment with stearic acid, W_{AB} is positive in air but negative in water. This result suggests natural peeling.

Table 2.4 Calculated Work of Adhesion for Composite Interfaces in the Presence and Absence of Water

Composite	W_{AB} (erg/cm^2)	W_{AB}^L (erg/cm^2)	$W_{AB} - W_{AB}^L$ (erg/cm^2)
Aluminum oxide/ polyethylene	120	−312	+432
Aluminum oxide/stearic acid/polyethylene	56.0	101.8	−45.8

Source: Ref. 13.

When stearic acid-treated aluminum is used, the adhesive work is positive even in water and the improvement in waterproofing is noticeable. In this case, a decrease in strength is expected in air as a result of the surface chemistry. However, due to the formation of a transition layer by the mutual diffusion of stearic acid membrane and polyethylene, the decrease in strength is not necessarily observable.

2.3   INTERACTION FORCE AT INTERFACE AND PEELING STRENGTH

To examine how the interaction force at the interface (one meaning of "adhesion force") reflects on the adhesion strength, we will review the peeling energy. In general, the peeling energy observed is known to be significantly larger than the interaction force at the interface. This is explained by the idea that the energy loss due to the change in the adhesion layer by peeling and the energy absorption near the broken surface greatly influence the peeling energy. Andrew et al. [14] defined the peeling energy, θ, by the equation

$$\theta = \theta_0 + \phi \tag{13}$$

and investigated the degree of each contribution. Here ϕ is a dissipation energy at the adhesion layer on peeling and ϕ_0 is the energy required for a crack to develop when $\phi = 0$.

As to the effect of ϕ, a temperature-rate dependence of the peeling strength is recognized, therefore the overlapping is naturally possible. Andrew et al. also compared the break strength (τ). In the adhesion of styrene-butadiene rubber (SBR) to various polymers shown in Table 2.5, the rate dependence of peeling and break energies is shown in Figure 2.4. It is elucidated from this figure that each curve is parallel to every other curve and that θ and τ are expressed by the equations

$$\theta_0 = \theta_0 F(R) \tag{14}$$

$$\tau = \tau_0 F(R) \tag{15}$$

Here F is a function and $R = C_{a\tau}$, the peeling rate. From equations (14) and (15),

$$\frac{\theta}{\theta_0} \equiv F(R) \equiv \frac{\tau}{\tau_0} \tag{16}$$

Table 2.5 Values of θ, the Intrinsic Adhesive Failure Energy θ_0, and the Thermodynamic Work of Adhesion W_a for Rubber Adhering to Various Substrates

Substrate	$\theta(mJ/m^2)$	$\theta_0(mJ/m^2)$	$W_a(mJ/m^2)$
FEPA	2.0×10^3	21.9	48.4
PCTFE	6.8×10^3	74.9	62.5
Nylon 11	6.5×10^3	70.8	71.4
PET	7.2×10^3	79.4	72.3
Plasma-treated FEPA	6.3×10^3	68.5	56.8
FEP C20	2.6×10^4	288	61.1
FEPA etched for			
10 s	7.8×10^4	851	68.0
20 s	1.1×10^5	1170	70.2
60 s	1.2×10^5	1290	69.8
90 s	1.5×10^5	1620	71.1
120 s	1.6×10^5	1780	71.1
500 s	2.2×10^5	2420	72.2
1000 s	1.8×10^5	1990	71.8

Source: Ref. 14.

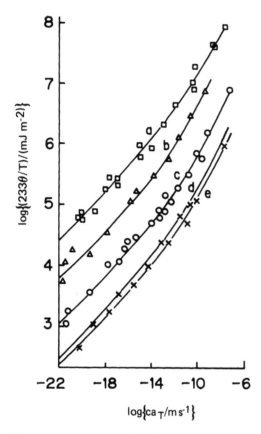

Figure 2.4   Adhesive failure energy against reduced rate of crack propagation at T_g for SBR with various substrates. a, τ for the adhesive; b, FEPA (etched 120 s); c, FEP C20; d, PET; e, nylon 11. (Ref. 14.)

τ and τ_0 are obtained by other methods and θ_0 is obtained by measuring θ. Also by using various liquids, the surface free energy of adhered materials and adhesives is determined to obtain the work of adhesion (W_a). The results are shown in Table 2.5.

From this, when the adhered material is FEPA through plasma-treated FEPA, W_a is almost equal to θ_0. This indicates that the interaction at the interface comes primarily from intermolecular forces. However, θ is almost 100 times as great as θ_0. This means that an energy of about 100-fold the value of the interfacial interaction force is consumed as a viscoelastic deformation energy of the adhesion layer.

Of course, the magnitude depends on the peeling temperature and its rate. But in FEPA etched by sodium naphthalenide, θ_0 becomes 20 to 30-fold the value of W_a. In other words, the interfacial interactions in this case include not only the intermolecular force, but also a stronger interaction force of a few 10-fold. This may be because double bonds are introduced into the FEPS surface by etching, chemical bonds are formed at the interface, and the percentage of cohesion break is increased. In other words, when a cohesion break occurs, θ becomes more than 1000-fold greater than W_a.

From these results, Schultz et al. [15] proposed the following equation for the peeling strength, where the dissipation factor [g(M)] related to the length of the molecular chain on the break face was considered in addition to the ideas of Andrew et al. [14]:

$$\theta = W_0 F(R) g(M) \tag{17}$$

Here θ, W_0, and $F(R)$ are respectively, a peeling energy, a bond energy at the interface (i.e., a physical interaction energy such as intermolecular forces), and a dissipation energy depending on the rate, similar to equation (14). In the case of interfacial peeling, $g(M) = 1$ gives a universal equation for peeling, irrespective of the break mode.

Furthermore, from electron microscopic measurements and those of the contact angle of the broken surface, Schultz et al. [15] confirmed that peeling of the adhered material between rubber and aluminum which had been subjected to pore-seal treatment by boiling water after cathodic oxidation in a sulfuric acid bath, was the result of an interfacial break. They obtained the peeling energy in liquid or in air and investigated the peeling mechanism using the equations that follow. When the peeling energies in air and in liquid are assumed to be θ and θ_L (W and W_L in the original paper), respectively, and the thermodynamic interaction forces are W_0^a and W_{0L}^a, respectively, $\theta = W_0^a f(R)$ and $\theta_L = W_{0L}^a f(R)$ are obtained. From this result, the following equations are derived:

$$\frac{\theta_L}{\theta} = \frac{W_{0L}^a}{W_0^a} \tag{18}$$

or

$$\frac{\theta_L - \theta}{\theta} = \frac{W_{0L}^a - W_0^a}{W_0^a} \tag{19}$$

or

$$\frac{\Delta \theta}{\theta} = \frac{\Delta W_0^a}{W_0^a}$$

(20)

$\Delta \theta / \theta$ is measured and $\Delta W_0^a / W_0^a$ can be calculated from the surface energies of the adhesive, the adhered material, and the peeling liquid.

Figure 2.5 shows the result of the rate dependence of peeling energy in air and in ethanol. It shows that the ratio between the two is constant. That indicates no rate dependence, and the value of the peeling energy in ethanol calculated from equation (19) shows good agreement with the observed value. The measured value ($\Delta \theta / \theta$) and the theoretical value ($\Delta W_0^a / W_0^a$) in other liquids are also shown in Table 2.6, where all of the theoretical and observed values show good agreement.

In other words, in this case, the interaction force at the interface is composed only of an intermolecular force; moreover, an ideal interface break occurs. In the case of cohesion break, if the peeling energies in air and in liquid are assumed to be θ and θ_L, respectively, $\theta = W_0^c f(R)$ and $\theta_L = W_{0L}^c f(R)$ are obtained similarly.

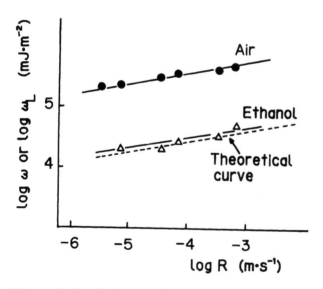

Figure 2.5   Influence of a liquid medium on the failure energy of an Al S-NBR assembly. (Ref. 15c.)

Table 2.6 Experimental and Theoretical Values of Changes in Peel Energy for Aluminum Substrate (%)

Liquid	Al S-NBR		Al S-SBR	
	$\dfrac{\Delta W}{W}$	$\dfrac{\Delta W_0^{\,a}}{W_0^{\,a}}$	$\dfrac{\Delta W}{W}$	$\dfrac{\Delta W_0^{\,a}}{W_0^{\,a}}$
Methanol	− 96	− 93	− 97	− 99
Ethanol	− 91	− 91	− 90	− 99
Butanol	− 93	− 89	− 92	− 99
Water	− 92	− 98	− 80	− 71

Source: Ref. 15c.

Since this case is a cohesion break, the interaction force at the interface is assumed to be W_0^c. We consider for W_0^c not only the intermolecular force, $W_0^c(\text{phys}) = 2\gamma_S$, but also the energy needed to break the C—C bond in the molecule, $W_0^c(\text{chem})$.

$$W_0^c = 2\gamma_S + W_0^c(\text{chem}) \tag{21}$$

Therefore, similar to equation (20),

$$\frac{\Delta\theta}{\theta} = \frac{\Delta W_0^c}{2\gamma_S + W_0^c(\text{chem})} \tag{22}$$

From this equation

$$W_0^c(\text{chem}) = \frac{\Delta W_0^c}{(\theta_L/\theta) - 1} - 2\gamma_S \tag{23}$$

is derived.

θ_L/θ is obtained from measurement of the peeling energies in air and in liquid, and ΔW_0^c and γ_S are obtained from the surface free energy. Therefore, $W_0^c(\text{chem})$ can be calculated. For example, an

Table 2.7 Determination of Cohesive Properties of Elastomers SBR and NBR in the Fracture Zone from Peeling Experiments

Elastomer	W_L/W exp. in ethanol	ΔW_0^c(theo) (mJ/m^2)	W_0^c (mJ/m^2)	W_0^c(phys) = $2\gamma_s$ (mJ/m^2)	W_0^c(chem) (mJ/m^2)
SBR	0.44	−52	93	60	33
NBR	0.18	−69	84	72	12

Source: Ref. 15b.

Table 2.8 Chemical Contribution to the Reversible Energy of Cohesion, W_0^c(chem (mJ/m^2), Calculated Using Various Network Models

Elastomer	Model of Lake and Thomas	Model of Bueche	Model of Flory and Rehner
SBR	50	33	42
NBR	35	22	29

Source: Ref. 15b.

aluminum surface treated by a mixed solution of phosphoric acid/ chromic acid is adhered to SBR or neoprene-butadiene rubber (NBR) and peeled to give a cohesion break.

Therefore, the peeling energy is measured in air and in liquid and W_0^c(chem is calculated from equation (12) to give the results shown in Table 2.7. The W_0^c(chem) values of SBR and NBR are calculated from this as 33 and 12 mJ/m^2, respectively. The chemical bond contributes only 35 and 14%.

Since W_0^c(chem) is a product of the number of broken molecular chains per unit break area (γ) and the dissociation energy of the $C-C$ bond (μ), it can be calculated if γ is known. Table 2.8 shows the results calculated using various cross-link models. These values seem to give a good agreement with the experimental values obtained in peeling experiments.

Recently, Furukawa [16] pointed out a lack of consideration of the physical nature of polymers in previous adhesion theory and proposed a new theory based on the pseudo-cross-link model shown in Figure 2.6. The peeling strength (f) is expressed by the following equation:

$$ f \propto \frac{\nu_e k_B Tv}{k''} \propto \nu_e k_B Tv \exp\left(\frac{E}{RT}\right) \tag{24} $$

where ν_e, k_b, k'', v, and T indicate a pseudo-cross-link in the adhesive, a variable constant of the pseudo-cross-link, a rate constant of flow, a peeling rate, and temperature, respectively. According to this theory, a WLF-type equation at low temperature and an Arrhenius equation at high temperature are claimed to give a good correlation with the observed values for the peeling.

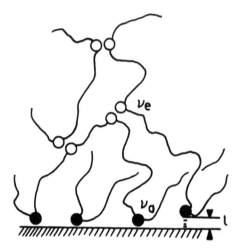

Figure 2.6 Schematic representations of pseudo-cross-link model. ν_e, pseudo-cross-link; ν_a, interfacial pseudo-cross-link. (Ref. 16.)

2.4   SURFACE CHEMICAL CHANGE OF THE ADHERED SURFACE AND ITS ADHESIVENESS

Conventional adhesion theory assumes that the surface of the adhered material does not change and that γ_S and its components γ_S^d, γ_S^p, γ_S^H, and so on, are constant. However, as expected from the fact that the polymer chain undergoes micro Brownian movement and that the additive, such as a plasticizer, undergoes blooming onto the surface over time, the nature of the polymer solid surface is considered to change depending on exterior circumstances.

Recently, it has become a matter of speculation whether the polar group on the surface inverts or moves into the interior as a result of very local movement of the polymer chain on the surface. For example, the fact that a polymer surface that has improved wettability via corona or glow discharge treatment loses its wettability over time is due to the inversion phenomenon of the polar substituent generated by the glow discharge treatment on the polymer surface [17].

Such an inversion phenomenon was investigated in view of measurements of the contact angle, with the results shown in Table 2.9 [18]. When film having an earlier increased contact angle is dipped in water, most of the polar substituents inverted into the interior of the film are reinverted to appear on the film surface, giving a contact angle of less than 10°.

Table 2.9 Contact Angles of Films Subjected to Plasma Exposure, Aging, and Water Immersion

Film	After exposure[a]		After aging[b]		After water immersion[c],[d]	
	θ_a	θ_r	θ_a	θ_r	θ_a	θ_r
Silicon	42.0	21.5	81.3	37.7	<10	<10
HDPE	<10	<10	68.2	18.5	<10	<10
PET	<10	<10	48.0	16.3	<10	<10
Polycarbonate	<10	<10	50.3	11.2	<10	<10

Source: Ref. 18.

[a] Ar, 20 ml/min[1], 6 min, 25 W (θ was determined 10 min after exposure).

[b] Air, 100°C, 12 min.

[c] 30°C, 6 days.

[d] Inverted bubble method.

When acrylamide was graft-polymerized on the surface of silicone resin, it was demonstrated by surface analysis via ESCA that even the graft chain moved into the resin [19]. Such a phenomenon is illustrated schematically in Figure 2.7. It is therefore assumed that a part of the polymer chain having polar substituents generated on the polymer surface rotates and the surface structure changes into a thermodynamically stable state. The driving force for such an inversion is a thermodynamic factor that makes the surface free energy as low as possible.

In the past, it was difficult using adhesion theory to explain why the same resin has varying degrees of adhesiveness depending on molding or treatment methods. Some of these phenomena can be elucidated in the light of polar inversion phenomena.

Table 2.10 shows the adhesiveness when urethane coating was added to the surface of various molded resins [20]. Even the same resin sometimes shows a large difference in adhesiveness, depending on the molding method. For instance, regardless of molding methods, low-density polyethylene (LDPE) or ethylene/vinyl acetate copolymer (EVA) having no polar group will not adher.

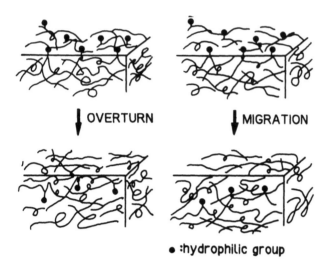

Figure 2.7   Schematic representations of the overturn and the migration of a polymer chain with hydrophilic groups.   (Ref. 18.)

A resin with polar substituents shows varying degrees of adhesiveness depending on the existing position of the polar group or the difference in chemical structure.   Therefore, in a random copolymer where the polar substituents are connected directly to the main chain, such as ethylene/methacrylic acid copolymer (EMAA) or saponified EVA, the difference in adhesiveness due to the molding method is small and the adhesiveness is excellent.

In a graft copolymer of maleic anhydride, where the carboxylic group exists in a form of a graft chain, the adhesiveness varies greatly depending on the molding method.   When a polyester film with polarity is used as a detached film, it shows a high level of adhesiveness.

A product from an injection mold or from a hot press mold employing a hydrophobic fluoro resin (Teflon) shows no adhesiveness. Furthermore, in blended polar and hydrophobic resins, the position of the polar group or the difference in chemical structure has an effect on the adhesiveness.   Especially when the graft chain is polar, a difference in molding method results in a big difference in adhesiveness.

Therefore, even when the polar group concentration is significantly lowered by the blend method, good adhesiveness to the surface of the material molded by a press method is shown using polyester film.   However, when the concentration of the polar group in

Table 2.10   Adhesiveness of Urethane Coating Film to Molding Sheet Surfaces[a]

| Polymer type | Method of sheet molding — Polymer contacting material during molding[b] | Hot press molding | | Injection |
		Polyester film	Teflon-TFE film	Chilled mold
Nonpolar Polymer	LDPE	0	0	0
	EVA (28% VA)	0	0	0
Polar polymer	Random Copolymer EMAA (12% MAA)	100	100	100
	EVA (14% VA)-90% saponified	100	100	100
	Graft Copolymer EVA (14% VA)-MAH (1.4%) graft	100	0	0
	HDPE-MAH (2.3%) graft	100	0	0
Blend of (nonpolar/polar) polymer	EVA (10% VA) EVA (14% VA) = 92/8	100	—	0
	EVA (10% VA) EVA (14% VA)-MAH (1.4%) graft = 96/4	100	—	0
	-MAH (1.4%) graft = 98/2	61	—	0
	EVA (10% VA) EMAA (15% MAA) = 90/10	0	—	0

Source: Ref. 20.

[a]Number of unpeeled tessellated incisions: Cut hundreds of tessellated incisions on coating film with a knife, then peel the incisions using adhesive tape, and count the number of unpeeled incisions (cross-cut test).

[b]EVA, ethylene/vinyl acetate copolymer; EMAA, ethylene/methacrylic acid copolymer; MAH, maleic anhydride.

the random copolymer is lowered, the material shows no adhesiveness using any molding method.

In the case of a graft copolymer of EVA (14% VA)/maleic anhydride (MAH 1.4% graft) adhesiveness again appears when the nonadhesive surface layer molded on Teflon film is eliminated. Furthermore, when the nonadhesive molded material is dipped in hot water at 95°C for 30 min, the surface tension increases again and adhesiveness appears. These phenomena mean that the segment containing the polar group moves on the solid surface and the solid surface is not always constant.

Furthermore, when a polar group is introduced into the main chain by random copolymerization, rotation of the polar group around the main chain is difficult and the effect of the molding method is barely observable. However, when a polar group is introduced by graft copolymerization, the polar group is easy to move and the surface characteristics change greatly during molding depending on the contact material. It seems important to pay careful attention to these points, especially when working with composite materials using polymers that have polar groups as the matrix.

REFERENCES

1.   P. Ehrburger and J. B. Donnet, Interface in Composite Materials, *Philos. Trans. R. Soc. London*, *A294*, 495 (1980).

2.   L. A. Girifalco and R. J. Good, A Theory for the Estimation of Surface and Interfacial Energies: I. Derivation and Application to Interfacial Tension, *J. Phys. Chem.*, *61*, 904 (1957).

3a. F. M. Fowkes, Additivity of Intermolecular Forces at Interfaces, *J. Phys. Chem.*, *67*, 2538 (1963).

3b. F. M. Fowkes, Attractive Forces at Interfaces, *Ind. Eng. Chem.*, *56*, 40 (1964).

4a. D. K. Owens, Some Thermodynamic Aspects of Polymer Adhesion, *J. Appl. Polym. Sci.*, *14*, 1725 (1970).

4b. D. K. Owens and R. C. Wendt, Some Thermodyanmic Aspects of Polymer Adhesion, *J. Appl. Polym. Sci.*, *14*, 1725 (1970).

5.   D. H. Kaelble and K. C. Uy, A Reinterpretation of Organic Liquid-Polytetrafluoroethylene Surface Interactions, *J. Adhes.*, *2*, 50 (1970).

6.   S. Wu, Polar and Nonpolar Interactions in Adhesion, *J. Adhes.*, *5*, 39 (1973).

7a. Y. Kitazaki and T. Hata, Wettability of Fluorine Substituted Polyethylenes, *J. Adhes. Soc. Jpn.*, *8*, 178 (1972).

7b. Y. Kitazaki and T. Hata, Estimation of the Extended Fowke's Equation and Surface Tension of Polymer, *J. Adhes. Soc. Jpn.*, *8*, 131 (1972).

8. Y. Ikada, *Fundamental and Application of Polymer Surface I*, Tokyo Kagaku Dozin Co., Ltd., Tokyo, 1986, p. 40.

9. A. Zosel, Adhesion and Tack of Polymers; Influence of Mechanical Properties and Surface Tensions, *Colloid Polym. Sci.*, *263*, 541 (1985).

10a. M. Imachi, Hot Melt Adhesion and Wettability Between Polyethylene and Other Polymers in the Vicinity of the Adhered Melting Point, *J. Appl. Polym. Sci.*, *34*, 2485 (1987).

10b. M. Imachi, Hot Melt Adhesion and Wettability of Polyethylene/Metal in the Vicinity of the Metal Melting Point, *J. Polym. Sci. Polym. Lett. Ed.*, *25*, 129 (1988).

11. K. Sumiya, T. Tani, K. Nakamae, and T. Matsumoto, Adhesion of the Vacuum-Deposited Cobalt Thin Films to Polymer Films, *J. Adhes. Soc. Jpn.*, *18*, 345 (1982).

12. A. Carre and J. Schultz, Polymer-Aluminum Adhesion: I. The Surface Energy of Aluminum in Relation to Its Surface Treatment, *J. Adhes.*, *15*, 151 (1983).

13. H. Schonhorn and H. L. Frisch, Environmental Aspects of Adhesion and Adhesive Joint Strength, *J. Polym. Sci.*, *11*, 1005 (1973).

14. E. H. Andrews and A. J. Kinloch, Mechanics of Adhesive Failure I, *Proc. R. Soc. London*, A*332*, 385 (1973).

15a. A. Carre and J. Schultz, Polymer-Aluminum Adhesion. II. Role of the Adhesive and Cohesive Properties of the Polymer, *J. Adhes.*, *17*, 135 (1984).

15b. A. Carre and J. Schultz, Polymer-Aluminum Adhesion: III. Effect of a Liquid Environment, *J. Adhes.*, *18*, 171 (1984).

15c. J. Schultz and A. Carre, Adhesion and Cohesion of Elastomers, *J. Appl. Polym. Sci. Appl. Polym. Symp.*, *39*, 103 (1984).

16. J. Furukawa, Theory of Adhesion Based on Pseudo-Cross-Link Model, *J. Adhes. Soc. Jpn.*, *23*, 407 (1987).

17. H. Yasuda and A. K. Sharma, Effect of Orientation and Mobility of Polymer Molecules at Surfaces on Contact Angle and

Its Hysteresis, *J. Polym. Sci. Polym. Phys. Ed.*, *19*, 1285 (1981).

18. Y. Ikada, T. Matsunaga, and M. Suguki, Overturn of Polar Groups on Polymer Surface, *Chem. Soc. Jpn.*, *6*, 1079 (1985).

19. B. O. Ratner, P. K. Weathersby, and A. S. Hoffman, et al., Radiation-Grafted Hydrogels for Biomaterial Application as Studies by the ESCA Technique, *J. Appl. Polym. Sci.*, *22*, 643 (1978).

20. E. Hirasawa and R. Ishimoto, Orientation of Polar Groups at Polymer Surface and Its Effects on Bondability, *J. Adhes. Soc. Jpn.*, *18*, 247 (1982).

3
Surface Modification of Matrix Polymer for Adhesion

3.1   INTRODUCTION

Composites are materials which have characteristics that cannot be obtained from a single material alone but which can be obtained by combining and synthesizing various materials that are different in nature (e.g., polymers, inorganic powders, metals, and oxides of them and their fibers) and configuration. An interfacial contact is thus inevitably between the different materials, and it is well known that the characteristics sought for in the material are strongly influenced by the bonding at the contact surface (interface). To derive full performance from the composites, many unknown problems concerning the nature of the interface between different materials, and the behavior and reaction at the interface, must be clarified and solved.

Conventionally, these interface problems are seen as a type of adhesion phenomenon and are often interpreted in terms of the surface structure of the bonded material. That is, surface factors such as wettability, surface free energy, the polar group on the surface, and the surface roughness of the material to be bonded are often discussed as means of improving the bonding strength. Recently, there has been work in which adhesion is understood as a volume phenomenon within a certain depth of material; these aspects of adhesion theory involve the diffusion theory of Voyutskii [1].

For the purpose of enhancing interaction at the interface between a filler and a matrix polymer of composite material, it is necessary to surface-modify the filler or the matrix polymer. This chapter deals with the adhesive properties of matrix polymers enhanced by the introduction of oxygen-containing groups such as —OH, —COOH, and >CO into matrix polymers to improve the wettability of the polymers and to form an active surface layer.

3.2   COMPLEXITIES OF ADHESIVE STRENGTH

The complexities of adhesive strength, which is affected by many factors, are central to an understanding of the subject matter of this book. In general, the adhesive property of a material is indicated directly by the adhesive strength. However, adhesive strength is not determined simply by the surface properties of the material to be bonded. The value of the adhesive strength is changed considerably, depending on conditions such as temperature, speed, and sample size.

In some cases, experimental results for two materials are compared in vain because there is only one difference between the measurement conditions applied for the test. Also, it is not always true that because a dry adhesive strength is higher, the adhered portion has greater water resistance.

Further, in some examples where there has been a thick adhesive layer or an adhesive layer with a low modulus of elasticity, the shear strength of the adhered portion decreases inversely with an increase in the peel strength. Difficulties involved in studies on adhesion are caused by the weak boundary layer (WBL) suggested to be present on the surface of a material. For instance, in an examination of variations in adhesive strength caused by plasma reactions, much attention should be given to the complexities of adhesive strength described above, because in plasma surface treatment, a variety of reactions occur (as described later), and there are many possibilities for complications in the reactions.

3.3   CHEMICAL TREATMENT OF POLYMERS TO
         IMPROVE WETTABILITY

When a polymer is soaked in a heavily oxidative chemical liquid and treated under suitable conditions, polar groups are introduced on the polymer surface and the surface characteristics are improved. Various types of chemicals, such as chromic anhydride/tetrachloroethane, chromic acid/acetic acid, chloratesulfuric acid, and potassium dichromate/sulfuric acid have been investigated as the heavily oxidative chemicals. Other methods of treatment, such as oxidation with NO, cycloalkyl chromate treatment, potassium permanganate treatment, sodium hypochlorite treatment, chlorosulfonation treatment, and flame treatment in the presence of halogen have been proposed. The most effective and common chemicals among these candidates is potassium dichromate/sulfuric acid. The surface of the polymer is heavily oxidized by nascent oxygen generated through the following reaction:

$$K_2Cr_2O_7 + 4H_2SO_4 \longrightarrow Cr_2(SO_4)_3 + K_2SO_4 + 4H_2O + 3O$$

(1)

The surface of polyolefine is activated by treating it with the liquid through the formation of polar groups such as $>C=O$, $-OH$, $-COOH$, and $-SO_3H$. Rasmussen et al. determined these polar groups qualitatively in detail [3]. In general, polyethylene is more sensitive than polypropylene, and low-density polyethylene is more sensitive than high-density polyethylene. The following mechanism for the formation of oxygen-containing polar groups has been proposed [4]:

(2)

The change in surface morphology is observed with an electron microscope. A lamella structure about 500 Å in height is observed on the surface of untreated high-density polyethylene but is not clearly observed on the surface of untreated low-density polyethylene. By chemical treatment, the surface of both polymers becomes rough, and holes of about 500 Å are formed on the surface. On the other hand, no lamella structure is observed on the surface of polypropylene, and when treated for 100 s the surface becomes rough (diameter 0.2 to 0.5 μm, height 0.1 μm). Use of a scanning electron microscope reveals no remarkable features before surface treatment, but there is a clear spherulite structure after treatment. This is caused by the fact that the amorphous portion between spherulites is more easily etched than is spherulite itself.

Figure 3.1 shows the change in surface wetting with treatment [4]. The contact angel (θ) of both types of polyethylene decreases with treatment, attaining about 60°C after 100 s but there is little change with further treatment. In the case of polypropylene, the contact angle (θ) decreases slightly after 5 s of treatment, but thereafter tends to increase. The decrease in contact angle in the

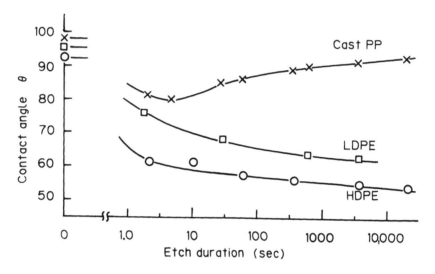

Figure 3.1 Contact angle measurements on etched polyolefin surfaces (advancing water droplets at 25°C). (Ref. 4.)

initial stage of treatment is believed to be due to the formation of
—OH groups, and the increase in contact angle with further treatment is probably due to the removal, during a water bath, of oxidized products that have formed on the surface.

Similar experimental results obtained by Briggs are shown in
Table 3.1 [5]. In the case of polyethylene (PE), the higher the
treatment temperature, the smaller the contact angle (θ) with water
becomes; the longer the treatment period, the smaller the contact
angle (θ) becomes. In the results with treatment at 70°C, when
there is overnight washing after treatment, the contact angle again
increases. This result is probably due to the removal of surface
layer formed during treatment. In the case of polypropylene (PP),
a tendency similar to that shown in Figure 3.1 is observed; treatment at 70°C for 6 hours results in a greater contact angle than that
obtained at 20°C for 1 min.

As mentioned above, the wettability of the surface is greatly improved by surface oxidation treatment, but adhesion behavior is
very complex, and a simple correlation between adhesion and quantity
of polar groups has not been found. For example, the following
varying results, which involved a contradiction between treatment
conditions and adhesion, were reported upon treatment of polyethylene with potassium dichromate/sulfuric acid: (1) the adhesion increases with increasing treatment period and temperature [6]; (2)
maximum adhesion is obtained at a certain intermediate treatment

Table 3.1 XPS Analytical Data[a], Contact Angles, and Joint Strength Data for Chromic Acid-Etched Polyethylene Surfaces

Etching conditions	C:S atomic ratio	O:S atomic ratio	Percent C atoms with SO₃H groups	Percent O (oxygen) (not in SO₃H groups) to total C	Advancing θ (deg)	Lap shear strength (MN/m²)	Failure type[b]
PE							
Untreated	—	—	—	0.25	98	0.55	I
1 min/20°C, normal wash	269	14.9	0.37	4.4	76	7.45	M
30 min/70°C, normal wash	80.0	12.2	1.25	11.5	66	7.58	M
6 h/70°C, normal wash	47.1	11.2	2.12	17.4	45	9.48	M
6 h/70°C, overnight wash	59.2	14.1	1.69	18.6	64	6.96	M
PP							
Untreated	—	—	—	0.25	92	0.28	I
1 min/20°C, normal wash	283 (223)	19.1 (19.0)	0.35 (0.45)	5.9 (7.3)	73	4.69	I and M
1 min/20°C, overnight wash	583 (307)	27.1 (19.4)	0.19 (0.33)	4.5 (5.3)	82	4.83	I and M
6 h/70°C, normal wash	261	15.9	0.38	4.8	97	11.2	M

Source: Ref. 5.
aData refer to an electron emission angle (θ) of 75°.
bI, apparent interfacial failure; M, failure of polyolefin film.
The values in parentheses mean similar data for other samples.

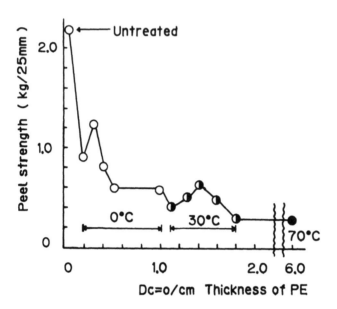

Figure 3.2   Effect of C=O content (optical density) in polyethylene on peel strength of aluminum plate/polyethylene/Al-foil laminate. (Ref. 9.)

period and temperature [7]; and (3) adhesion decreases with a further increase in treatment period [8].

Nakao et al., who paid special attention to >C=O among the polar groups, determined the relation between the quantity of the >C=O formed and peel strength, and obtained the results shown in Figure 3.2 [9]. In the figure it is obvious that maximum adhesion is obtained in the region where small amounts of >C=O such as are not detectable by infrared spectrometry are formed. This result suggests that the surface layer of polymer becomes oxidized and subject to deterioration, and both the molecular weight and the cohesive force are decreased. Correspondingly, the adhesion is decreased when more >C=O is formed in amounts detectable by infrared spectrometry.

It is therefore desirable that treatment conditions be set such that polar groups form on the surface layer of polymer without deterioration. At the same time, it should be noted that the presence of water during adhesion causes a decrease in adhesion. This fact suggests that the decrease is probably due to bonding between >C=O on the polymer surface and water molecules and that this bonding inhibits interaction between polymer and adhesive.

For polyesters and polycarbonates treatment methods include the use of chromic acid [10] and surface amination [11] (the same treatment as that used for polyethylene). For example, when polyester film is treated with a primary aliphatic amine at 80 to 150°C, the polymer chains are broken to form amino groups through the following reaction:

$$\text{www} \overset{\overset{O}{\parallel}}{C}\text{-}\langle O\rangle\text{-}\overset{\overset{O}{\parallel}}{C}\text{-COCH}_2\cdot\text{CH}_2\text{OC}\text{-}\langle O\rangle\text{-}\overset{\overset{O}{\parallel}}{C}\text{ www} \quad + \text{ RNH}_2 \qquad (3)$$

$$\longrightarrow \text{ www } \overset{\overset{O}{\parallel}}{C}\text{-}\langle O\rangle\text{-}\overset{\overset{O}{\parallel}}{C}\text{-NHR} + \text{HO}\cdot\text{CH}_2\text{CH}_2\text{O } \overset{\overset{O}{\parallel}}{C}\text{-}\langle O\rangle\text{-}\overset{\overset{O}{\parallel}}{C}\text{ www} \qquad (4)$$

Polycarbonates react with amine to form urethane groups and phenolic hydroxyl groups, and this reaction occurs at a relatively low temperature.

$$\text{www O}\text{-}\langle O\rangle\text{-}\overset{\overset{\text{CH}_3}{|}}{\underset{\underset{\text{CH}_3}{|}}{C}}\text{-}\langle O\rangle\text{-O}\overset{\overset{O}{\parallel}}{C}\text{O}\text{-}\langle O\rangle\text{-}\overset{\overset{\text{CH}_3}{|}}{\underset{\underset{\text{CH}_3}{|}}{C}}\text{-}\langle O\rangle\text{-} \quad +\text{RNH}_2 \qquad (5)$$

$$\longrightarrow \text{ www O}\text{-}\langle O\rangle\text{-}\overset{\overset{\text{CH}_3}{|}}{\underset{\underset{\text{CH}_3}{|}}{C}}\text{-}\langle O\rangle\text{-OH} + \text{RNHCO}\text{-}\langle O\rangle\text{-}\overset{\overset{\text{CH}_3}{|}}{\underset{\underset{\text{CH}_3}{|}}{C}}\text{-}\langle O\rangle\text{-O www} \qquad (6)$$

Adhesion is improved because polar groups are formed as described above. Also, the adhesion properties of propylene/methyl-1,4-hex-adiene copolymer are increased considerably by surface treatment with ozone, aqueous $KMnO_4$ solution, or concentrated H_2SO_4 [12].

3.4 MODIFICATION BY CORONA
DISCHARGE TREATMENT

Corona discharge treatment is widely practiced in industry as a
process for the activation of substrate surfaces as required to improve their adhesive properties. It is used, for example, for pretreatment of polyolefin films in lamination processes [13] and for
pretreatment of paper and paperboard in extrusion coating processes
[14]. Various mechanisms have been proposed for the improvement
of the adhesive properties of polyethylene by corona discharge treatment, some attributing it to electric formation [15], others to hydrogen bonding [16].

Corona discharge treatment results in the formation of high-
polarity functional groups such as carbonyl at the polymer surface.
Figure 3.3 presents ESCA analyses of Polyethylene treated by corona
discharge [17]. The spectrum of $C_{1}s$ shown in Figure 3.3(a)

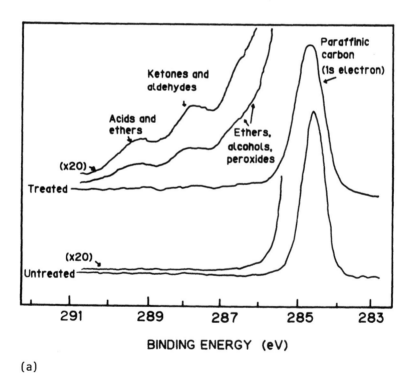

(a)

Figure 3.3   ESCA spectra of carbon (1s) photoline of untreated and
treated PE surfaces. (Ref. 17.)

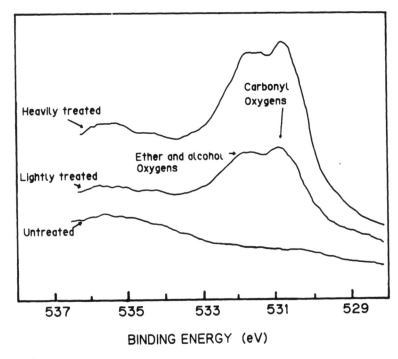

BINDING ENERGY (eV)

(b)

Figure 3.3 (Continued)

consists of peaks of carbons corresponding to functional groups of ethers, alcohols, peroxides, ketones, aldehydes, acids, and esters which are not detected in untreated material. Chemical shifts related to the bonding conditions of C_1s are +4.1 eV for $R*COOH$, +3.9 eV for $R*COO-R$, +3.2 eV for $R*CHO$, +3.1 eV for $R-C*O-R$, +3.3 eV for $R-C*O-NHR$, +1.5 eV for $R*CH_2OH$, +1.6 eV for $RC*H_2---O-R$, and 0 eV for $R-C*H_2-CH_2-R$. By means of these values, the peaks in Figure 3.3(a) can be correlated with various functional groups.

Oxygen-containing functional groups formed by corona discharge treatment can be estimated from ESCA spectra of O_1s. ESCA spectral measurements on oxygen are shown in Figure 3.3(b). Oxygen atoms contained in ethers, alcohols, and carbonyl are detected as treatment proceeds.

An infrared absorption spectrum around 1720 cm^{-1} is shown in Figure 3.4 [15]. As one would expect, absorption due to carbonyl

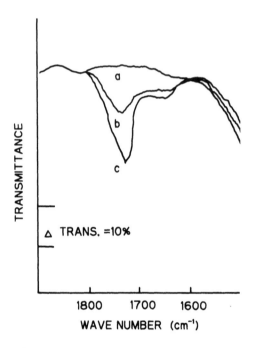

Figure 3.4   MIR infrared spectra of PE: a, control or film treated in nitrogen for 30 min; b, film treated in oxygen for 20 s at 25°C; c, film treated in an ozone spin reactor for 30 min.   (Ref. 15.)

appears in an oxygen atmosphere. Nitrogen-containing functional groups such as $-ONO_2$ and $-NO_3^-$ are detected in addition to these carbonyls [18].

Figure 3.5 illustrates measurements of self-adhesive properties and critical surface tension of polyethylene treated by corona discharge in various gas atmospheres [15]. Self-adhesive properties are little changed by treatment in hydrogen, whereas they are improved after treatment for several seconds in oxygen, air, carbon dioxide, or 31% hydrogen-containing nitrogen.

Furthermore, no effects on wettability are brought about by treatment in an hydrogen atmosphere, whereas an improvement is seen in oxygen, air, carbon dioxide, or nitrogen. Figure 3.6 shows observations on adhesion to, and wettability by ink of, polyethylene samples treated in nitrogen, helium, and argon atmospheres. Adhesion is seen to increase in all atmospheres, whereas wet tensile strength tests do not indicate substantial effects in helium and argon atmospheres [19].

For poly(ethylene terephthalate), self-adhesive properties are improved by corona discharge treatment. This improvement is considered to be a result of the formation of polar functional groups triggered by oxygen atoms or ultraviolet light produced during the discharge process [20]:

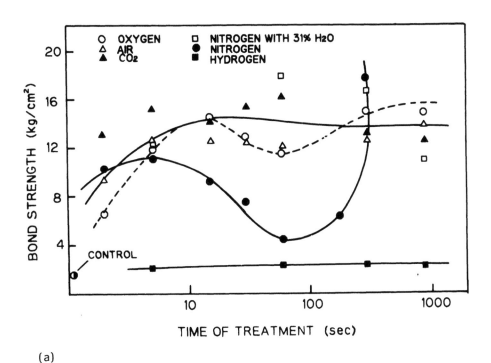

Figure 3.5   (a) Variation of bond strength with time of treatment of PE treated in corona discharges of various gases.   (Ref. 15.)

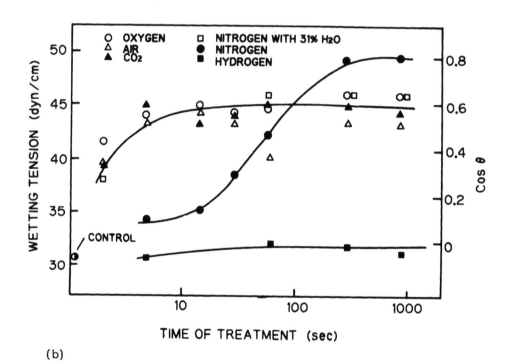

(b)

Figure 3.5   (b) Changes in wetting tension and water contact with time of treatment of PE surfaces treated in corona discharges of various gases. (Ref. 15.)

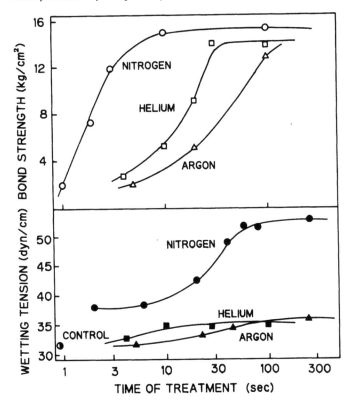

Figure 3.6   Effect of time of corona discharge treatment in nitrogen, helium, and argon gas on ink adhesion and wetting tension. (Ref. 19.)

3.5   MODIFICATION BY ULTRAVIOLET IRRADIATION

As a process for surface modification, ultraviolet irradiation treatment has various advantages: (1) reaction occurs at ordinary temperature and pressure, (2) selective reaction is possible, and (3) light energy can be focused on the surface of material. In most cases, modification is carried out (1) by introducing functional groups to the material surface either by applying ultraviolet light to oxidize the material surface or by allowing the material to contact a gas or sensitizer to cause a photochemical reaction, or (2) by allowing ultraviolet irradiation graft polymerization to occur at the material surface.

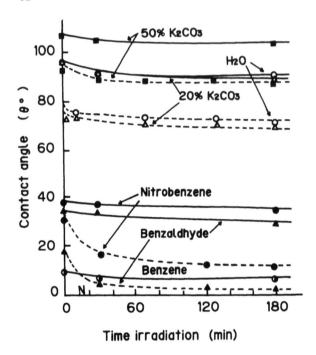

Figure 3.7   Change of contact angle for polyethylene irradiated
with ultraviolet light (—), or treated with chromium sulfate mixed
acid (---).   (Ref. 21.)

     For polyethylene irradiated with ultraviolet light, Figure 3.7
illustrates the relation of irradiation time to contact angle for various liquids.   The dashed line in Figure 3.7 represents measurements
of polyethylene samples treated with chromium sulfate mixed acid.
Wettability is not as high in polyethylene samples irradiated with
ultraviolet light as in those treated with chromium sulfate mixed
acid [21].   Changes in critical surface tension (γ_c) are not as
large as those for untreated polyethylene, as seen in Table 3.2.
     Figure 3.8 illustrates the effects on adhesive properties.   Ad-
hesive strength increases with increasing degree of surface treatment when an epoxy adhesive is used, whereas it does not show a
large increase in the case of a neoprene adhesive [22a].   Shown in
Figure 3.9 for comparison are changes in the adhesive properties
of a polyethylene sample treated with chromium sulfate mixed acid.
For poly(ethylene terephthalate), ultraviolet irradiation acts to
change the chemical structure at the surface making it hydrophilic
[equations (10) through (12)] [20].
     Table 3.3 shows measurements on adhesion to copper foil.   Among
the various surface treatment processes, ultraviolet irradiation serves

Table 3.2 γ_c of Surface-Treated PE

Surface treatment		γ_c (dyn/cm)
Untreated		30
Ultraviolet	30 min	34
irradiation	180 min	35
Chromic acid/	1 min	38
sulfuric acid	60 min	42
	180 min	44
Discharge	0.01 mmHg	35
treatment	0.05 mmHg	47.5
	0.1 mmHg	55.0
	1 mmHg	42.0
	10 mmHg	38.5

Source: Ref. 21.

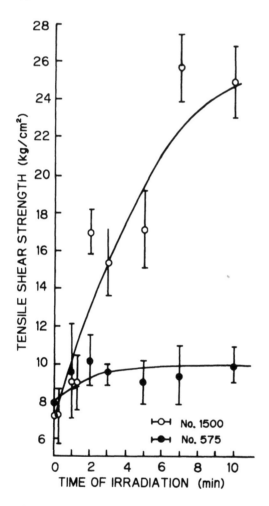

Figure 3.8   Change of tensile shear strength for polyethylene with time of irradiation of ultraviolet light: adhesive 1500, blend of epoxy and polyamide resins; adhesive 575, chloroprene.   (Ref. 22a.)

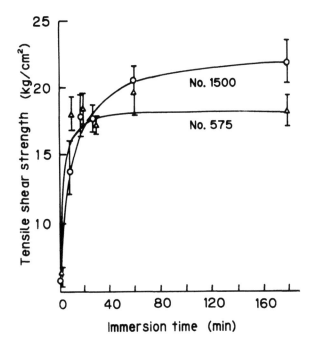

Figure 3.9   Change of tensile shear strength for polyethylene with time of immersion in chromic acid mixtures (immersion temperature 25 ± 2°C).   (Ref. 22a.)

most effectively to improve adhesive properties [23]. Functional groups such as $-NO_2$ and $-Cl$ are added to the polyethylene surface when irradiation by ultraviolet light is applied in gases such as Cl_2, NO, and NO_2 [24]. As an application of such reactions, an attempt was made to enhance the adhesion of propylene-ethylene copolymers to paints through irradiation with ultraviolet light in Cl_2 gas followed by treatment with phenyl lithium or alkali [25].   To modify a polyethylene surface, carboxylic acid is produced at the surface by introducing double bonds through irradiation in acetylene [equation (13)] followed by photooxidation [26].

$$PE \xrightarrow[HC\equiv CH]{UV} \quad \overset{|}{\underset{HC=CH_2}{\frown}} \quad \xrightarrow[O_2]{UV} \quad \overset{|}{\underset{COOH}{\frown}} \qquad (13)$$

In addition, there are many surface modification processes that use triplet sensitizers to permit oxidation reactions. In a typical

Table 3.3   Relation Between Peel Strength and Surface Treatment

Method of treatment	Condition			Peel strength (kg/cm)
No treatment	–	–	–	0.04
NaOH treatment	30% NaOH	80°C	3 min	0.27
UV irradiation treatment	In air	8 cm	30 s	0.05
	–	–	1 min	0.05
	–	–	2 min	0.35
	–	–	5 min	0.40
	–	–	10 min	0.45
	–	–	30 min	0.40
	–	5 cm	5 min	0.52
	–	–	10 min	0.53
	–	–	30 min	0.59
	–	2 cm	5 min	0.43
	–	–	10 min	0.45
	–	–	30 min	0.49
	100 mmHg	8 cm	10 min	0.35
	10 mmHg	–	10 min	0.35
	1 mmHg	–	10 min	0.40
	0.1 mmHg	–	10 min	0.40
Ti treatment	Titanium-acetylacetonate	–	–	0.05
	Tetra-isopropyltitanate	–	–	0.04
Hydrazine treatment	Immersion, at room temperature for:	–	–	–
	–	–	5 min	0.25
	–	–	10 min	0.25
	–	–	30 min	0.25
	–	–	60 min	0.15
	–	–	120 min	0.10

Source: Ref. 22b.

process, polyisocyanate is applied on a polyolefin together with a sensitizer such as benzophenone and then irradiated with ultraviolet light. As shown in equation (14), the sensitizer has an oxidizing effect to produce hydroxyl groups over the polymer surface. These hydroxyl groups finally react with isocyanate to provide a functional primer [27,28].

$$PE \xrightarrow[O_2]{UV} \qquad OH \xrightarrow{R(NCO)_3} \qquad OCNHR(NCO)_2 \qquad (14)$$
$$\overset{\|}{O}$$

In some cases, adhesive properties can be improved by ultraviolet light irradiation alone [29].

3.6 MODIFICATION BY PLASMA TREATMENT

The physical properties of a composite material are dominated by adhesion between interfaces in a composite material, as pointed out repeatedly in this book. A variety of attempts have been made to enhance the adhesive properties of the surfaces of reinforcing materials. One such attempt, the utilization of plasma, is a typical dry treatment method [30]. Radiation and photochemical methods, which also are of dry type, are seldom used at present, due to the fact that the number of reactive modified groups per unit surface area of material produced by radiation or photochemical treatment is very small compared to the number produced by plasma treatment. Plasma surface treatment is already used by industry to improve the painting performance of plastic motor vehicle bumpers and the printability of films. In this section, plasma treatments to improve the adhesive properties of reinforcing materials for composites and for general polymeric materials are described.

3.6.1 Plasma Reaction of Polymer

The plasma utilized for polymer treatment is more properly called non equilibrium low-temperature plasma [31]. In this plasma, the temperature of the whole gas is not the same as that of the electrons present in the gas. In particular, the plasma generally indicates the state of a substance in which positively and negatively charged particles coexist and move freely, and thus a substance in which the electric charge is neutral. On the other hand, in low-temperature plasma for polymer treatment, relatively few electrons and ions are present in the gas. This is different from the high-temperature plasma of the sun and in nuclear fusion. Accordingly, the temperature of the whole low-temperature plasma gas is so low that the degree of heat resistance of the material being treated requires little attention. Only a small number of electrons present in the gas have high energies.

It is known that the energy of electrons (e_f) present in low-temperature plasma is in the range of 1 to 10eV. This energy causes molecules of gas A to be ionized and excited. As a result, radicals and ions are produced, as expressed in the following formulae:

$$\text{Ionization:} \qquad A + e_f \longrightarrow A^+ + 2e \qquad (15)$$

$$\text{Excitation:} \qquad A + e_f \longrightarrow A^* + e \qquad (16)$$

$$\text{Radical dissociation:} \qquad A^* \longrightarrow A_1^{\cdot} + A_2^{\cdot} \qquad (17)$$

$$\text{Luminescence:} \qquad A^* \longrightarrow A + h\nu \qquad (18)$$

$$\text{Electron addition:} \qquad A + e \longrightarrow A^- \qquad (19)$$

The activated particles react with the polymeric material so that polymeric radicals are produced on the surface layer of the material. The polymeric radicals eventually cause the surface layer to be oxidized, cross-linked, or decomposed. On the other hand, A·'s are produced from molecules of the gas and are polymerized, so that the resultant polymers of A coat the surface of the material. A scheme that shows such a process is shown in Figure 3.10. As listed in Table 3.4, these reactions are utilized for various modifications of polymeric surfaces.

An outline of the plasma reaction, which is more closely associated than any other reaction with improvement in the adhesive properties of polymeric materials, is described below. It should be noted that the effects of the plasma reaction depend to a great extent on the type of plasma reactor, the oscillation frequency, the high-frequency output, the gas type, the flow rate and pressure of the gases, the treatment time, the position of the samples, and so on.

3.6.2 Cross-linking

It has been reported that when a polymeric material is plasma-treated in an inert gas such as helium or argon, cross-linkages are introduced into the surface layer of the material. Schonhorn has developed a method of enhancing the adhesive properties of materials by utilization of cross-linkage formation [32]. He has named this method "cross-linking by activated species of inert gases" (CASING). Figure 3.11 shows a suggested mechanism for the cross-linking reaction of polyethylene. Figure 3.12 summarizes the bonding test results, which indicate that the bonding strength is improved.

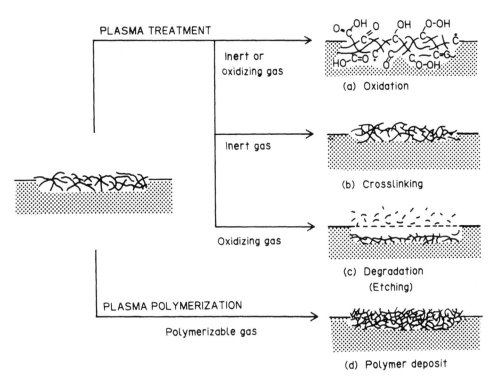

PLASMA TREATMENT

Inert or oxidizing gas

(a) Oxidation

Inert gas

(b) Crosslinking

Oxidizing gas

(c) Degradation (Etching)

PLASMA POLYMERIZATION

Polymerizable gas

(d) Polymer deposit

Figure 3.10 Schematic of plasma modifications of polymer surface.

Table 3.4 Interaction of Plasmas with Polymer Surface

Gas	Reaction	Application example
Inert	Radical formation Cross-linking	Graft copolymerization Surface protection
Oxidizing	Oxidation Degradation	Wettable surface Etching
Functional	Substitution	Introduction of group
Polymerizable	Deposit	Ultrathin film coating

$$He^* + RH \longrightarrow R\cdot + H\cdot + He$$

$$He^* + R_1R_2 \longrightarrow R_1\cdot + R_2\cdot + He$$

$$H\cdot + RH \longrightarrow H_2 + R\cdot$$

$$R\cdot + R_1\cdot \longrightarrow RR_1$$

$$R\cdot + R_2\cdot \longrightarrow RR_2$$

$$R\cdot + R\cdot \longrightarrow RR$$

Figure 3.11   Mechanism of polyethylene cross-linking by plasma treatment in inert gas.   (Ref. 32.)

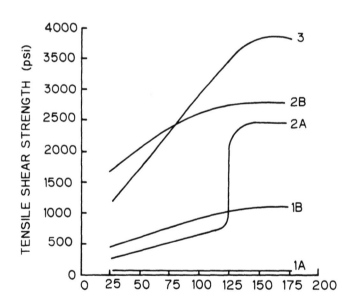

Figure 3.12   Effect of inert-gas plasma on the tensile shear strength of lap shear composites: 1A, untreated Teflon film; 1B, Teflon film + 10 min CASING; 2A, untreated polyethylene (Marlex 5003) film; 2B, Marlex 5003 + 10 s CASING; 3, aluminum/epoxy resin/aluminum. (Ref. 32.)

It has been suggested that the bonding strength is enhanced because the cross-linkages strengthen the surface layer and prevent a WBL from being formed. However, there are diverse opinions as to whether or not the enhancement of adhesive properties is based on cross-linkage formation alone. This is because reactions other than cross-linking occur to such a significant degree that contributions of these reactions to the enhancement of adhesive properties cannot be neglected, as described below.

3.6.3 Oxidation and Decomposition of Polymer Surfaces

When polymers are plasma-treated under relatively mild conditions, oxidation reactions generally occur. As a result, oxygen atoms can be introduced [33] into polymeric surfaces having high resistance to chemical reactions, even for silicone, fluoro resins, and polyethylene, resulting in the formation of carbonyl, carboxyl, hydroxyl, and aldehyde groups and carbon-carbon double bonds, as shown in Figure 3.13.

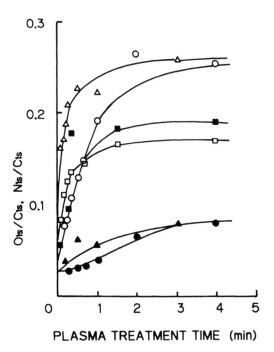

Figure 3.13 O_{1s}/C_{1s} and N_{1s}/C_{1s} intensity ratios for the PE film exposed to different plasmas of 11.5 W. $\triangle$, air; $\circ$, Ar; $\square$, N_2 for O_{1s}/C_{1s}. $\blacktriangle$, air; $\bullet$, Ar; $\blacksquare$, N_2 for N_{1s}/C_{1s}. (Ref. 33.)

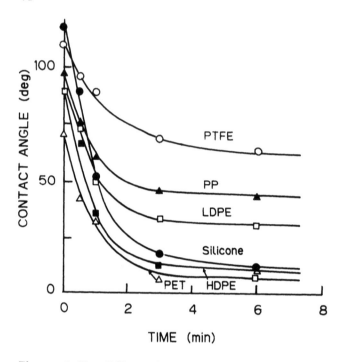

Figure 3.14 Effect of plasma exposure time on contact angle. ○, PTFE; ●, silicone; □, LDPE; ■, HDPE; ▲, PP; △, PET. (Ref. 34.)

Also, in the case of plasma treatment in an ambient inert gas, polar groups such as those formed by oxidation reactions are produced. Materials plasma-treated in an inert gas always come into contact with air at some time following treatment. Trapped radicals generated by the plasma treatment react with oxygen in the air. Accordingly, polar groups, although relatively few in number, are produced in the same manner as in plasma treatment in an oxidative gas. Needless to say, these polar groups will noticeably affect the bonding strength. Schonhorn may not have considered such polar groups.

When polar groups such as carboxyl groups are introduced into the hydrophobic surface of a material, its wettability is enhanced. Figure 3.14 shows a typical example [34]. However, wettability deteriorates (i.e., the contact angle to water increases) with increased time standing in air, as shown in Figure 3.15. Wettability is recoverable by dipping the material in water (Table 3.5).

Many studies indicate that enhancing wettability by plasma treatment results in high bonding strengths. It is presumed that the improved wettability improves the spread of an adhesive on the surface of a substrate, which in turn improves the bonding strength. Figures 3.16 and 3.17 [35] show two examples of the results of studies by Hall et al. in which they plasma-treated various polymeric materials in helium and oxygen and then bonded the materials to each other using an epoxy adhesive.

Figure 3.16 shows the relation between shear strength and plasma treatment time for various PEs, and Figure 3.17 that of poly(4-methyl-1-pentene) bonded to poly(vinyl fluoride). In both cases, the bonding strength increases rapidly for an initial relatively short plasma treatment time. Thereafter, the bonding strength levels off or decreases. A change in shear strength R due to plasma treatment is expressed satisfactorily by the formula

$$R = \frac{1}{a + bt}$$

where t designates plasma treatment time, and a and b are constants.

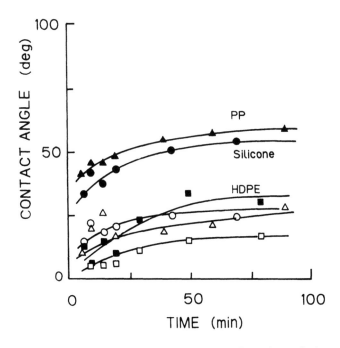

Figure 3.15   Contact angles as a function of the aging time after plasma exposure (plasma exposure time, 6 min). ○, ●, silicone; □, ■, HDPE; △, ▲, PP. Open marks, receding contact angle; solid marks, advancing contact angle. (Ref. 34.)

Table 3.5   Contact Angles of Films Subjected to Plasma Exposure, Aging, and Water Immersion

Film	After exposure[a]		After aging[b]		After water immersion[c],[d]	
	θ_a	θ_r	θ_a	θ_r	θ_a	θ_r
Silicone	42.0	21.5	81.3	37.7	<10	<10
HDPE	<10	<10	68.2	18.5	<10	<10
PET	<10	<10	48.0	16.3	<10	<10
Polycarbonate	<10	<10	50.3	11.2	<10	<10

Source: Ref. 34.

[a]Ar, 20 ml/min, 6 min, 25 W (θ was determined 10 min after exposure).

[b]Air, 100°C, 12 min.

[c]30°C, 6 days.

[d]Inverted bubble method.

As can be understood from the foregoing examples, when polar groups are introduced into the surface of a material by plasma oxidation, the dry bonding strength is increased notably compared with that of untreated material.   However, in some cases the in-water bonding strength is not enhanced significantly because the polar groups are so attractive to water that the interface has a tendency to be attacked by water and easily peeled.

One of the difficulties with plasma treatment of polymers is that the polymer is decomposed by the plasma reaction.   After a long period of plasma treatment or with treatment at an elevated high-frequency output, decomposition will have proceeded to a considerable degree and the polymeric material will be degraded or etched.   Degradation of polymers by plasma is positively utilized for removal of polymeric stains from the surfaces of inorganic materials such as metals and glass and for the incineration of analytical samples.

Also, in the case of polymeric materials, a WBL formed on the material can be eliminated by plasma treatment, which leads advantageously to enhancement of the adhesive properties.   In some cases it is hypothesized that an increase in the bonding strength by plasma treatment is based on removal of the WBL.   However, this is difficult to prove experimentally.

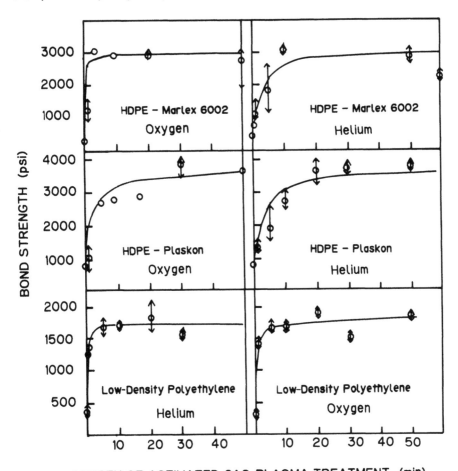

Figure 3.16   Bond strength of polyethylene versus length of exposure to activated gases.   (Ref. 35.)

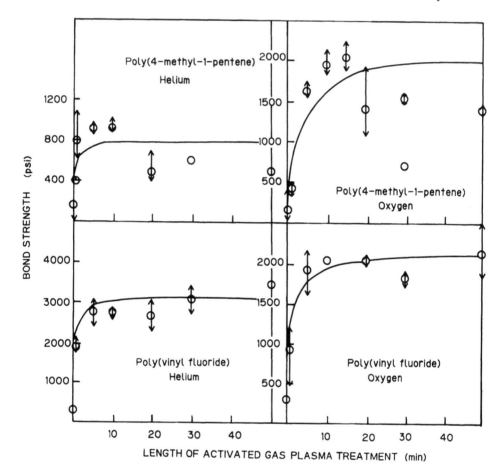

Figure 3.17 Bond strength of poly(4-methyl-1-pentene) and poly-(vinyl fluoride) versus length of exposure. (Ref. 35.)

3.6.4 Formation of Roughened Surface Layer

The structure of the roughened surface layers of some polymers appears to enhance significantly the polymers' adhesive properties. Recently, to roughen polymer surfaces, etching techniques have been widely employed. Dry surface treatment of the polymers is most desirable, followed by chemical surface treatment using acids, alkalis, and so on. The low-temperature plasma generated using a gas conditioned into a low-pressure state is especially characteristic, in that it activates the substratum by its physical action, that is, by the collision of high-energy electrons and ions against the surface substantially without the heat of the plasma acting on the substratum. Low-temperature plasma is utilized for a variety of surface treatments, such as surface modification or production of polymeric materials, plasma etching of electronic materials, and so on.

A sputtering-etching technique utilizing a low-pressure glow discharge process is effective in fine roughening of the surface of polymers. However, use of this technique to form a roughened surface layer is not available for all polymers. For example, the sputtering-etching technique is useful for polytetrachloroethylene-polyhexachloroethylene copolymers (PEP), polytetrachloroethylene-perfluoroalkoxyl copolymers (PFA), poly(methyl methacrylate) (PMMA), polyoxymethylene (POM), and others. On the other hand, for poly(vinylidene fluoride) (PVDF) and polytetrafluoroethylene-ethylene copolymers (ETFE), the surface roughening technique is not available.

In the surface roughening of polymers by etching techniques, it is not only by etching itself that the surfaces are roughened. In addition, gaseous monomers generated and excited by the etching are made to adhere to the tips of protrusions formed on the etched surface, and are polymerized. At a result, protrusions grow simultaneously with the etched surface formed. For PVDF and ETFE, which are difficult to surface-roughen, as described above, it has

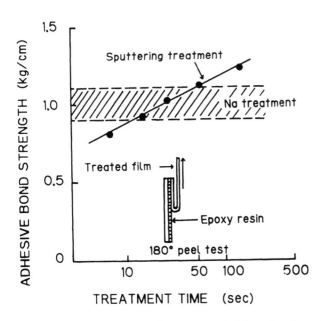

Figure 3.18   Effect of sputter etching treatment on the adhesive strength of PTFE film.   (Ref. 23.)

been proposed that the difficulty is caused by the fact that the monomers generated by etching are not made to adhere to, and do not grow on, protrusions formed on the etched surface [36].

As for the surface treatment of polytetrafluoroethylene, the adhesive properties of the polymer can be enhanced significantly by treating the surface with the plasma of an inert gas such as argon accelerated in an electric field for etching. Polytetrafluoroethylene surface-treated by the sputtering-etching method to be bonded to a steel sheet with an epoxy resin adhesive was subjected to the 180° peel test. Figure 3.18 shows the bond strength between the polymer and the sheet [23]. Generally, polytetrafluoroethylene surface-treated by a corona discharging or CASING method exhibits improved surface wettability and adhesive properties. However, sufficient strength for practical use cannot be obtained [37].

The ESCA spectra of surface-etched polytetrafluoroethylene show that sputtering-etching surface treatment scarcely changed the chemical properties of the polytetrafluoroethylene at the surface [38,39]. This is significantly different from polytetrafluoroethylene surface-treated with sodium-ammonia and polarized at the surface. Based on this fact, it has been suggested that the needle-shaped protrusions formed by sputtering-etching surface treatment have a physical anchoring effect which enhances polymer's adhesive properties. According to the hypothesis, this fact contributed to the smaller reduction in polymer adhesive properties by ultraviolet irradiation. Sputtering-etching techniques are employed widely to roughen the surfaces of polymers such as polyimide [40].

REFERENCES

1. T. Hata, Physiology of Adhesion, *J. Soc. Rubber Chem., Jpn.*, 45, 883 (1972).

2. Adhesion Society of Japan, *Handbook of Adhesion and Adhesive.* ed., Nikkan Kogyo Shinbun Ltd., Tokyo, 1971.

3. J. R. Rasmussen, E. R. Stedronsky, and G. M. Whitesides, Introduction, Modification, and Characterization of Functional Groups on the Surface of Low-Density Polyethylene Film, *J. Am. Chem. Soc.*, 99, 4736, 4746 (1977).

4. P. Blais, D. J. Carlsson, G. W. Csullog, and D. M. Wiles, The Chromic Acid Etching of Polyolefin Surfaces and Adhesive Bonding, *J. Colloid Interface Sci.*, 47, 636 (1974).

5. D. Briggs, D. M. Brewis, and M. B. Konieczo, X-Ray Photoelectron Spectroscopy Studies of Polymer Surfaces, *J. Mater. Sci.*, 11, 1270 (1976).

6. W. H. Shrader and M. J. Bodner, Adhesive Bonding of Poly-ethylene, *Plast. Technol.* (Dec.), 988 (1957).

7. M. J. Bodner and W. J. Powers, Adhesive Bonding of the Newer Plastics, *Plast. Technol.*, (Aug.), 721 (1958).

8. A. A. Berlin, S. F. Bulacheva, and Yu. L. Morozov, Modifica-tion of the Properties of Polyethylene by Surface Oxidation, *Sov. Plast.*, 4(10), 4 (1962).

9. K. Nakao and M. Nishiuchi, Studies on Adhesion of Polyolefin, Part 5: Effect of Potassium Dicromate Sulfuric Acid Solution Treatment on Peel Strength of Polyethylene, *J. Adhes. Soc. Jpn.*, 2, 239 (1966).

10. Mobey Chemical Co., Coating of Plastic Polycarbonates, Belg. Pat. 613,962 (1962).

11. J. R. Caldwell, Surface-Treating Polyester Films and Fibers with Primary Amino Compound, U.S. Pat. 2,921,828 (1960).

12. S. Kitagawa, I. Okada, and R. Itoh, Reactive Polyolefine, *Polym. Bull.*, 10, 196 (1983).

13. H. E. Wechsberg and J. B. Webber, Surface Treatment of PE Film by Electrical Discharge, *Mod. Plast.*, 36, 101 (1959).

14. R. E. Greene, Flame or Electrical Discharge Priming of Paper Substrate in High-Speed Extrusion Coating, *Tappi*, 48, 80A (1965).

15. C. Y. Kim, J. Evans, and D. A. I. Goring, Corona-Induced Autohesion of Polyethylene, *J. Appl. Polym. Sci.*, 15, 1365 (1971).

16. D. K. Owens, Mechanism of Corona-Induced Self-Adhesion of Polyethylene Film, *J. Appl. Polym. Sci.*, 19, 265 (1975).

17. H. L. Spell and C. P. Christenson, Surface Analysis of Corona-Treated Polyethylene, *Tappi*, 62, 77 (1979).

18. H. A. Willis and J. I. Zichy, *Polymer Surface*, ed. D. T. Clark and W. J. Feast, John Wiley & Sons, Inc., New York, 1978, p. 289.

19. M. Stradal and D. A. I. Goring, The Effect of Corona and Ozone Treatment on the Adhesion of Ink to the Surface of Polyethylene, *Polym. Eng. Sci.*, 17, 38 (1977).

20. D. K. Owens, The Mechanism of Corona and Ultraviolet Light-Induced Self-Adhesion of Polyethylene Terephthalate Film, *J. Appl. Polym. Sci.*, 19, 3315 (1975).

21.   T. Tsunoda, K. Chiba, and M. Fukumura, Change of Wetting by Surface Treatment of Polyethylene, *J. Chem. Soc. Jpn. Ind. Chem.*, *72*, 2451 (1969).

22a.   T. Tsunoda, Y. Oba, K. Chiba, and M. Fukumura, Adhesion and Surface Properties of Pretreated Polymers, *Nippon Kagaku Kaishi*, 659 (1978).

22b.   T. Tsunoda, Interfacial Phenomena and Adhesion, *J. Adhes. Soc.*, *Jpn*, *18*, 511 (1982).

23.   T. Tsunoda, Y. Oba, and M. Fukumura, Surface Properties of Ultraviolet Irradiated Polyethylene Terephthalate, *Kobunshi Ronbunshu*, *35*, 229 (1978).

24.   J. F. Kinstle and S. L. Watson, Jr., Photoassisted Modification of and Grafting to Polyethylene, *Polym. Sci. Technol.*, *10*, 461 (1977).

25a.   Mitsubishi Kasei Co. Ltd., Surface Treatment of Polyolefin, Jpn. Pat. (unexam.) S-53-147,771 (1978).

25b.   Mitsubishi Kasei Co. Ltd., Surface Treatment of Polyolefin, Jpn. Pat. (unexam.) S-54-77 (1979).

26.   H. Kimura and H. Nakayama, Surface Treatment of Plastics, *Color Mater. Jpn.*, *54*, 149 (1981).

27.   R. A. Bragole, Factors Affecting the Adhesion of Paints to Non-polar Plastics and Elastomers, *J. Elastomers Plast.*, *6*, 213 (1974).

28.   C. D. Storms, Functional Primers for Plastics, *Plast. Des. Process.*, *17*, 57 (1977).

29.   H. Schonhorn and F. W. Ryan, Surface Crosslinking of Polyethylene and Adhesive Joint Strength, *J. Appl. Polym. Sci.*, *18*, 235 (1974).

30.   S. Kaplan and P. Rose, Plasma Surface Treatment of Plastics, 1542, SPI, Proc. 46th Annual Tech. Conf, & Exib., Atlanta, (1988).

31.   H. V. Boening, *Plasma Science and Technology*, Cornell University Press, Ithaca, N.Y., 1982.

32.   R. H. Hansen and H. Schonhorn, A New Technique for Preparing Low Surface Energy Polymers for Adhesive Bonding, *J. Polym. Sci. Polym. Lett. Ed.*, *4*, 203 (1966).

33.   M. Suzuki, A. Kishida, H. Iwata, and Y. Ikada, Graft Copolymerization of Acrylamide onto a Polyethylene Surface Pretreated with a Glow Discharge, *Macromolecules*, *19*, 1804 (1986).

34. Y. Ikada, T. Matsunaga, and M. Suzuki, Overturn of Polar Groups of Polymer Surface, *Nippon Kagaku Kaishi*, 1079 (1985).

35. J. R. Hall, C. A. L. Westerdahl, M. J. Bodnar, and D. W. Levi, Effect of Activated Gas Plasma Treatment Time on Adhesive Bondability of Polymer, *J. Appl. Polym. Sci.*, *16*, 1465 (1972).

36. S. Yamamoto, H. Tabata, K. Sasa, and T. Moriuchi, Sputter Etching Mechanism of Polymers by RF Glow Discharge, *Appl. Phys. Jpn.*, *53*, 727 (1984).

37. S. Yamamoto, Surface Treatment Technique of Plastic Films, *J. Plast. Process Technol. Soc. Jpn.*, *9*, No. 2, 9 (1982).

38. K. Hatada and H. Kobayashi, Surface Modification of Organic Polymers by Low-Temperature Plasma, *Surf. Sci. Jpn.*, *5*, 408, (1984).

39. G. Ke-Cheng and Z. Shao-Hua, Plasma Treatment on Polytetra-fluoroethylene and the Adhesion Property, 1555, SPI. Proc. 46th Annual Tech. Conf. & Exhib., Atlanta, (1988).

40. F. D. Egitto, F. Emmi, and R. S. Horwath, Plasma Etching of Organic Materials: I. Polyimide in O_2--CF_4, *J. Vac. Sci. Technol.*, *B3*, 893 (1985).

4

Surface Modification of Matrix Polymer by Graft Polymerization and Its Effect on Adhesion

4.1 INTRODUCTION

Graft polymerization is useful for introducing actively polar groups on a polymer surface. Various methods, including catalysis, oxidation, radiation, photoirradiation, and plasma polymerization have been developed for graft polymerization. To be suitable for the surface modification of polymer, such a method must take into account the following: (1) a heterogeneous reaction is apt to take place since the base polymer is a solid; (2) graft polymerization on non-polar polymer is possible; and (3) graft polymerization can take place only on the surface of stem polymer. Based on these factors, plasma treatment and photoirradiation and radiation graft polymerization are most suitable.

To modify the polymer surface effectively, the polymerization reaction should be limited to the surface itself as much as possible. Therefore, a monomer that will not penetrate or diffuse into a graft layer is desirable. Examples of suitable monomers include acrylamide, acrylonitrile, methacrylic acid, and acrylic acid.

The introduction of polar groups by graft polymerization is often used for nonpolar polymers, which have inherently poor surface wettability. Typical such polymers are the polyolefine polymers such as polyethylene and polypropylene and fluorine-containing polymers such as polytetrafluoroethylene.

4.2 PLASMA TREATMENT POLYMERIZATION

4.2.1 Plasma Polymerization

In electric discharging in ambient gaseous molecules containing elements such as fluorine, silicon, nitrogen, and hydrocarbons,

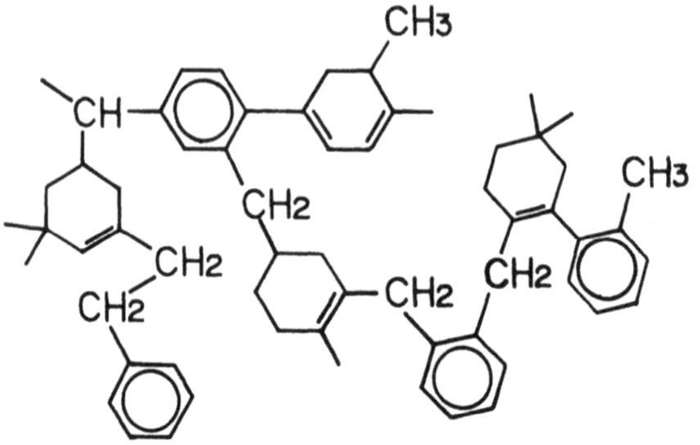

Figure 4.1   Model for chemical structure of toluene polymer produced by plasma polymerization. (Ref. 1.)

polymeric substances are often produced. In selected conditions, the polymer is formed into a thin film less than 1 μm in thickness which adheres tightly to the surfaces of glasses and metals as well as to plastics. This process is called "plasma polymerization film coating." The thin film has no pinholes, and the gas barrier performance is high. It has been suggested that the thin film has a highly cross-linked structure, as shown in Figure 4.1 [1]. Although the polymer is produced from hydrophobic molecules, the contact angle to water becomes small with continuing plasma polymerization (Figure 4.2), probably because the oxygen in the air reacts with radicals trapped in the film [2].

The adhesive properties of a substrate coated with a plasma polymerization film are often improved remarkably. Figure 4.3 shows the experimental results obtained when acetylene monomers are plasma-polymerized on polytetrafluoroethylene (PTFE) and then aluminum is bonded to the PTFE by use of an epoxy adhesive. As shown in Figure 4.3, high tensile shear strengths are obtained when the flow rate of the gas is low. The tensile shear strength increases as the output increases, as shown in Figure 4.4. Table 4.1 lists the maximum tensile shear strength values. Yasuda et al. recommend that plasma polymerization be carried out at the proper W/FM value to obtain a high tensile shear strength [3]. Here W designates a high-frequency output, M the molecular weight of a gas, and F the gas flow rate.

Generally, when a bonded part comes into contact with water, the bonding strength is reduced. This is one of the greatest shortcomings of bonding methods that utilize an adhesive. Inagaki et al. have indicated that a plasma-polymerization coating method gives

higher lap-shear strengths than those obtainable with a plasma oxidation treatment method (Table 4.2) [4]. Further, they have found
that it takes considerably longer to peel the adhered material from
the bonded part in contact with water when a plasma-polymerization
coating with methane, ethylene, or a similar compound is utilized.
Figure 4.5 illustrates the experimental results. In the case of an
acetylene plasma-polymerization film coating, the adhered material
is peeled off comparatively easily, as shown clearly in Figure 4.5.
In this case, an especially large number of polar groups may be
formed on the surface of the polymerization film.

4.2.2   Plasma-Induced Graft Polymerization

As mentioned above, carbonyl and hydroxyl groups and double
bonds are produced when polymers are plasma-treated. As shown
in Figure 4.6, peroxide groups are also formed [5]. By utilization
of the peroxide group, graft polymerization is allowed to proceed
as illustrated in Figure 4.7. Figure 4.8 shows typical experimental

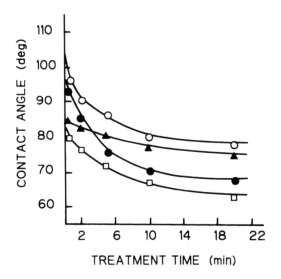

Figure 4.2   Dependence of contact angle with water of films coated
with plasma-polymerized acetylene on glow discharge treatment time.
Glow discharge conditions: pressure, 0.10 torr; acetylene flow rate,
0.21 mmol/min; discharge current, 15 mA; discharge voltage, 750 V.
o, PTFE film; ▲, PVC film; •, PE film; ▫, PVF film   (Ref. 2.)

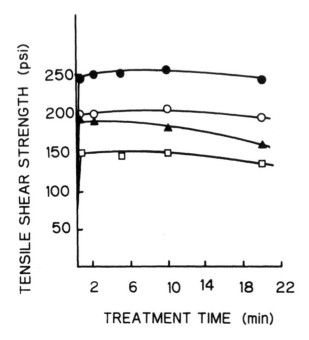

Figure 4.3   Dependence of tensile shear strength of the composite
aluminum/epoxy adhesive/acetylene glow-discharge-treated PTFE/epoxy
adhesive/aluminum on glow discharge treatment time at different
acetylene flow rates.   Glow discharge conditions: discharge current,
15 mA; discharge voltage, 750 V.   ●, 0.21 mmol/min acetylene flow
rate; ○, 0.46 mmol/min acetylene flow rate; ▲, 0.76 mmol/min acetylene flow rate; □, 1.89 mmol/min acetylene flow rate.   (Ref. 2.)

results obtained by graft polymerization of acrylamide monomer to
PE [5].   Graft polymerization of the water-soluble monomer makes
the PE surface hydrophilic to a high degree, and it never returns
to being as hydrophobic as is the surface of virgin PE.   The grafted
surface is different from that of PE that has simply been plasma-
treated.   It is probable that such grafting utilizing plasma enhances
the adhesive properties of polymeric materials.

4.2.3   Ion Implantation Methods

The methods of treating polymeric surfaces described above utilize
oxidation, decomposition, cross-linking of the polymer, plasma polymerization to form an organic thin film, and plasma-induced graft
polymerization of monomers utilizing radicals or peroxides.   An ion

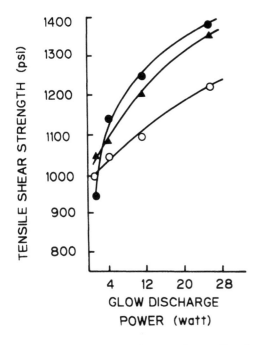

Figure 4.4 Dependence of tensile shear strength of the composite aluminum/epoxy adhesive/acetylene glow-discharge-treated PE/epoxy adhesive/aluminum on glow discharge power. Glow discharge condition: acetylene flow rate, 0.21 mmol/min. ○, 0.5 min treatment time; ▲, 5.0 min treatment time; ●, 20.0 min treatment time. (Ref. 2.)

Table 4.1 Maximal Tensile Shear Strength of the Composite: Aluminum/Epoxy Adhesive/Acetylene Plasma-Treated Film/Epoxy Adhesive/Aluminum[a]

Film	Film width (mm)	Tensile shear strength (psi)	
		Untreated	Treated
PE	0.10	90	1250
PVF	0.03	10	1450
PVC	0.20	900	1450
PTFE	0.20	10	250

Source: Ref. 2.
[a]Glow discharge condition: see Figures 4.3 and 4.4. Tensile shear strength was determined according to ASTM D-1002-64.

Table 4.2   Lap-Shear Strength[a]

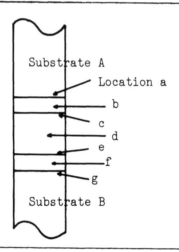

Construction of adhesive joint[b]		Lap-shear strength[c] (kg/cm^2)			
Substrate A	Substrate B	Uncoated	CH_4 (2000)[d]	$CH_2{=}CH_2$ (2000)[d]	$CH{=}CH$ (2060)[d]
PE	PE	5.8_9(b)	$24._2$(b)	$24._4$(b)	$20._7$(b)
PE	Al	$16._6$(b)	$47._6$(f)	$49._5$(f)	$48._2$(b)
PE	ST	$28._5$(b)	$45._1$(f)	$47._8$(f)	$46._5$(b)
T	T	2.4_7(b)	$17._9$(b)	$17._8$(b)	$17._8$(b)
T	Al	$15._8$(b)	$36._8$(f)	$40._4$(f)	$41._6$(b)
T	ST	3.8_0(b)	$38._0$(f)	$39._8$(f)	$40._3$(b)

Source: Ref. 4.

[a]Operational conditions: flow rate, 2.0 cm^3(STP)/min; current of ac power, 250 mA.

[b]PE, polyethylene; T, polytetrafluoroethylene; Al, aluminum, ST, stainless steel.

[c]Symbol in parentheses defined as follows illustrates the location of failure: a: interface between substrate A and plasma polymer; b: cohesive failure of plasma polymer layer on A side; c: interface

Table 4.2   (Continued)

between plasma polymer on A side and adhesive; d: cohesive failure of adhesive; e: interface between plasma polymer on B side and adhesive; f: cohesive failure of plasma polymer layer on B side; g: interface between substrate B and plasma polymer layer.

^{d}Film thickness (Å).

SUBSTRATE (A)	PLASMA FILM	GLUE	PLASMA FILM	SUBSTRATE (B)	SHEAR STRENGTH (kg/cm^2)
PE	(Ar)	G	(Ar)	PE	
PE	CH_4	G	CH_4	PE	
PE	CH_2CH_2	G	CH_2CH_2	PE	
PE	CHCH	G	CHCH	PE	
PE	(Ar)	G	(Ar)	AL	
PE	CH_4	G	CH_4	AL	
PE	CH_2CH_2	G	CH_2CH_2	AL	
PE	CHCH	G	CHCH	AL	
PE	(Ar)	G	(Ar)	ST	
PE	CH_4	G	CH_4	ST	
PE	CH_2CH_2	G	CH_2CH_2	ST	
PE	CHCH	G	CHCH	ST	

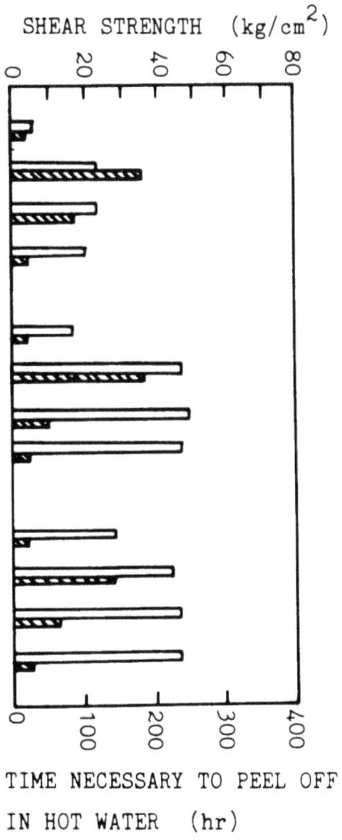

SHEAR STRENGTH

TIME NECESSARY TO PEEL OFF IN HOT WATER

TIME NECESSARY TO PEEL OFF IN HOT WATER   (hr)

Figure 4.5   Lap shear strength and deterioration in hot water as a function of material and gas used for polymerization. PE, polyethylene; Al, aluminum; ST, stainless steel; (Ar), argon etching; CH_4, plasma polymer from methane; CH_2CH_2, plasma polymer from ethylene CHCH, plasma polymer from acetylene. (Ref. 4.)

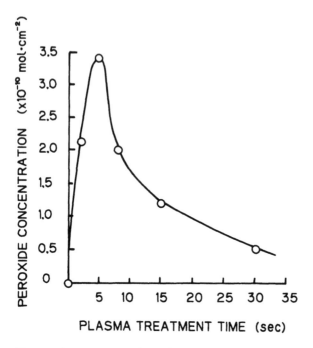

Figure 4.6   Formation of peroxides on the PE film exposed to argon plasma.   (Ref. 5.)

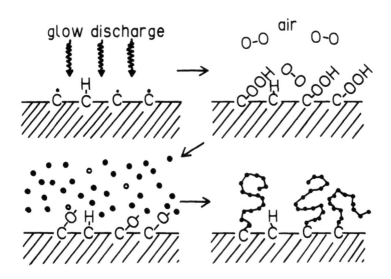

Figure 4.7   Schematic view of graft copolymerization of a water-soluble monomer onto the surface treated with glow discharge.   ●, monomer; ○, reducing agent.

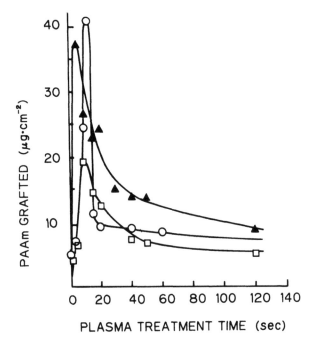

Figure 4.8 Dependence of the graft amount on plasma exposure time for the PE film exposed to different plasmas of 11.5 W followed by graft copolymerization. o, Ar; △, O_2; □, H_2; ●, N_2. (Ref. 5.)

implantation method in which accelerated ions are injected into the surface layer of a base material can also be used to enhance adhesive properties. In ion implantation an interlocking effect is obtained in the interface on the atomic level. For example, Yasuda et al. injected C^+ ions into platinum, which has no adhesive properties in relation to polymers, and methane molecules were plasma-polymerized [6]. They have found that methane polymerization film is not peeled off the platinum surface. It is suggested that the plasma polymer is adsorbed to the C^+ ions injected into the platinum surface.

Recently, a glow discharge plasma has been used to make composite elements form covalent bonds with each other prior to lamination. As a result, the composite's transverse and shear strengths increased 2.5-fold compared with those of conventionally processed composite of the same composition. Both chemical bonding of fiber and matrix and control of the surface energies of the two materials contribute to the strengthening effect. Plasma processing improves the adhesion strength of a phenolic bismaleimide/graphite fiber thermoset composite by a factor of 3. The strength of a graphite fiber/thermoplastic resin composite is augmented 2- to 2.5-fold [7a]. Further, oxidation of

polyethylene by corona discharge and the subsequent graft polymerization of acrylamide were reported by Iwata et al. [7b]. From the
X-ray photoelectron spectroscopic analysis, the corona treated polyethylene film surface and optical microscopy on the cross section of
the grafted film revealed the graft polymerization to be limited to a
very thin surface region.

4.3   RADIATION POLYMERIZATION

Since radioactive rays have much greater energy than that of ordinary
chemical bonds, irradiation with them serves to effect various modifications of polymers by causing such reactions as ionization, excitation,
and radical formation, which result in scission or three-dimensional
cross-linking of polymer chains. If the polymer radicals formed are
used to polymerize monomers of other species, properties of the newly
produced polymeric structures will be imparted semipermanently to the
original polymer. In addition, processes using radiation have various
advantages not available in conventional processes, such as the following: (1) catalysts are not required; (2) reactions can be performed
as close to ordinary temperatures; and (3) posttreatment is easy.

Irradiation may cause scission of polymer chains (degradation)
or the formation of three-dimensional linkages (cross-linking). Which
of these two reactions will occur depends on the structure of the
polymer and irradiation conditions. From the relation of these reactions to the heat of polymerization of monomers, it may be inferred
that polymers with high heats of polymerization tend to cross-link,
as shown in Table 4.3 [8]. It is also reported that in relation to
thermal stability, polymers capable of 100% decomposition are of the
"degradation type," while those capable of 30% or less decomposition
are of the "cross-linking type" [9].

With respect to the relation of irradiation conditions to cross-
linking or degradation, oxygen in the irradiation atmosphere has a
strong influence; for example, it causes polypropylene, which tends
to cross-link in a vacuum, to degrade in air. This arises from the
fact that peroxides, produced by irradiation in an oxygen-containing
atmosphere, suffer acid decomposition [10]. Similar phenomena are
seen in polystyrene [11] and poly(vinyl chloride) [12]. For polytetrafluoroethylene, degradation is rapid in air but becomes very low in
an oxygen-free atmosphere [13].

In general, irradiation acts to change the number of unsaturated
groups in polymers. In polyethylene, for example, vinyl and vinylidene groups disappear while transvinylene groups increase in number under irradiation [14]. Conjugated double bonds are formed in
poly(vinyl chloride) as a result of dehydrochlorination. However,
the number of double bonds tends to decrease in rubbers, which
originally contain many double bonds.

Table 4.3   Correlation of Polymer Properties with Irradiation Effects

Direction of irradiation effect	Polymer	Heat of polymerization (kcal/mol monomer)	Monomer yield (wt%)
Cross-link	Ethylene	22	0.025
	Propylene	>16.5	2
	Methyl acrylate	19	2
	Acrylic acid	18.5	–
	Styrene	17	40
Degrade	Methacrylic acid	15.8	–
	Isobutylene	13	20
	Methyl methacrylate	13	100
	α-Methyl styrene	9	100

Source: Ref. 8.

4.3.1 Application to Surface Modification and Effect on Adhesion

A typical method for surface modification by irradiation is to apply radiation to material to trigger graft polymerization. Two major techniques are simultaneous irradiation and preliminary irradiation. The latter may be performed either in air or in vacuum. As seen in Figures 4.9 and 4.10, grafting results in a sharp decrease in contact angle when preliminary irradiation with gamma or electron rays is carried out on a polyethylene film to cause acrylamide-grafting polymerization at the surface [15]. Graft polymerization takes place at low temperatures when Fe^{2+} is added to the reaction system as a reducing agent [16]. Observations on the optimum concentration of Fe^{2+} are presented in Figure 4.11. Adhesion of polyethylene to aluminum foil is increased when acrylic acid is grafted by preliminary irradiation with electron rays [17].

Figure 4.12 shows the relation of acrylic acid content to peel strength. The strength increases with increasing acrylic acid content, clearly demonstrating the polarization effect caused by grafting. Furthermore, polymers with a high dose of radiation show high adhesion strength at low temperatures, while those with a low dose show high adhesion strength at high temperatures. This is attributed to the difference in the molecular structure of the graft polymer. When the dose is high, the molecular weight of branches is low and its number is large. When the dose is low, on the other hand, the branches

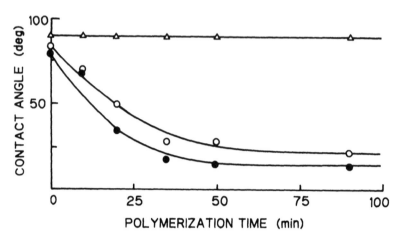

Figure 4.9   Change of contact angle of PE tubes by graft copolymer-ization (preirradiation in air to a dose of 7.0×10^5 rad).   ○, LDPE; ●, HDPE; △, nonirradiated LDPE.   (Ref. 15a.)

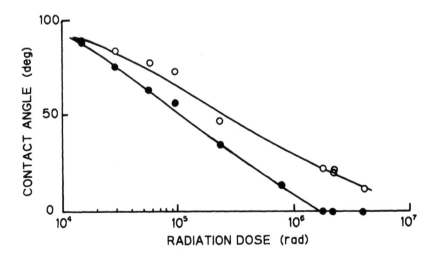

Figure 4.10   Dependence of water contact angle on radiation dose (polymerization for 1 h   at 50°C in the absence of Fe^{2+}).   ○, LDPE; ●, HDPE.   (Ref. 15b.)

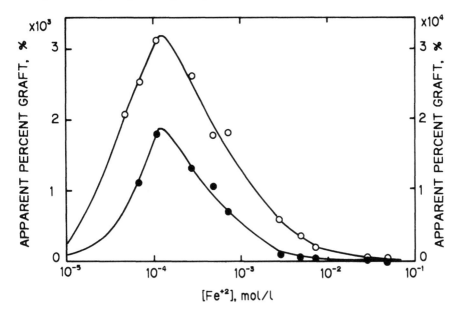

Figure 4.11 Dependence of apparent percent graft on Fe^{2+} concentration (irradiation to 4.5×10^6 rad, polymerization for 15 h at 30°C). ○, LDPE; ●, HDPE. (Ref. 15b.)

have high molecular weight but are small in number. In general, it is believed that molecular motion is relatively inactive at low adhesion temperatures to allow polymers with many branches to increase in adhesion activity, while molecular motion becomes active at high temperatures to permit polymers with long branches to work effectively.

As seen in Table 4.4, adhesive properties are improved when methyl acrylate chains grafted to polyethylene through polymerization by gas-phase simultaneous irradiation [18–20] are hydrolyzed into acrylic acid [21]. The surface produced by this process consists of graft layers and cross-linked homopolymer layers. The cross-linked homopolymer layers play an important role in increasing the peel strength, as indicated in Figure 4.13 [22]. As seen from Figure 4.14, however, the weather resistance of adhesive bonds is extremely small even if such layers exist and the initial peel strength is high [23]. This is considered to arise mainly from the removal of uniform homopolymer layers as a result of the photooxidation that takes place in the presence of water. This is demonstrated clearly in Figure 4.15, which indicates that the peel strength suffers only a slight

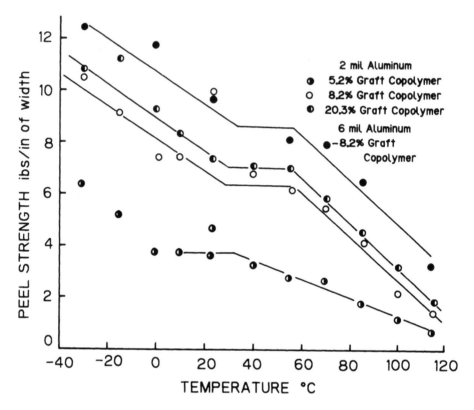

Figure 4.12   180° Peel adhesion versus temperature.   (Ref. 17.)

Table 4.4   T-Peel Strength of Treated PE Joints Consisting of
PE-Epoxy-PE[a]

| | T-peel strength (average−maximum−minimum) (kg/25 mm) | |
Treatment	LDPE	MDPE
Control	<0.1	<0.1
Hydrolyzed MA graft	>30[b]	>40[b]
Acid solution, 30 min, 70°C	2.1−7.4−0.6	4.1−20−1.7
O_2 plasma, 5−30 min, 25−50 W	6.0−38[b]−3.7	5.0−20−2.3
He plasma, 5−30 min, 75−100 W	0.4−0.8−0.1	2.9−5.3−2.2

Source: Ref. 21.
[a]Bonded with adhesive S-2.
[b]PE adherend failure.

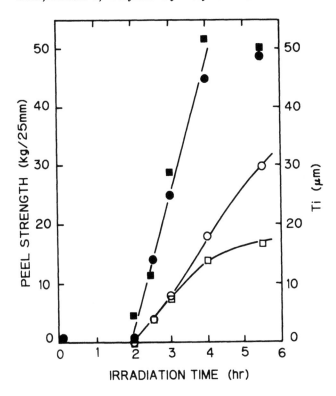

Figure 4.13 Peel strength and thickness of isotropic layer (Ti) of saponified surface grafts versus irradiation time. PE sheets (3.0 mm thick) were grafted and then saponified: •, peel strength, low-density PE; ■, peel strength, medium-density PE; o, Ti, low-density PE; □, Ti, medium-density PE. (Ref. 22.)

change if an oxidation stabilizer and an ultraviolet-light absorbent are added [23].

In Figure 4.16, the effect of the adhesion of polyethylene and polypropylene on which unsaturated carboxylic acid is graft polymerized to nylon is shown, and it is clear that adhesion is greatly improved by modification with carboxylic acid [25]. Polytetrafluoroethylene is a typical polymer that is difficult to adhere because of its extremely strong cohesive force, but the adhesive strength is greatly increased by graft polymerizing glycidyl methacrylate through radiation graft polymerization, as shown in Table 4.5 [25]. In another example, polytetrafluoroethylene on which methyl acrylate is graft polymerized and which is hydrolyzed exhibits good adhesion, as shown in Figure 4.17 [26].

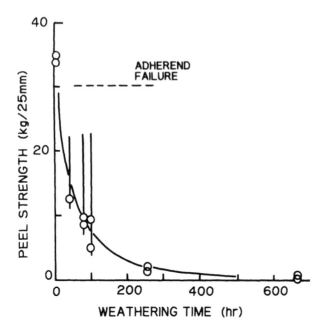

Figure 4.14   Changes in adhesive bondability of a grafted surface
with accelerated weathering. A γ-ray-induced graft of a carbon
black-containing low-density PE (2.0 mm thick) was used. (Ref. 23.)

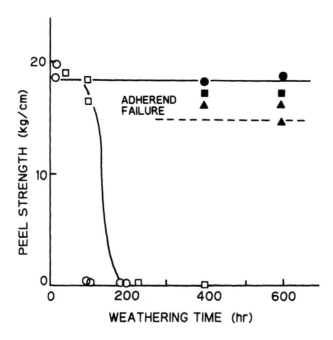

Figure 4.15

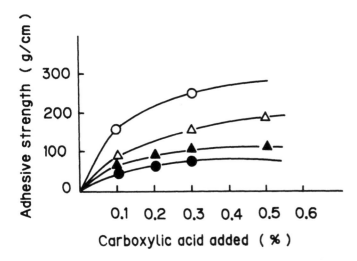

Figure 4.16 Relation between adhesive strength and unsaturated carboxylic acid added (%). Adhesive condition: 240°C, 4 kg, 5 s. ●, HDPE- acrylic acid; ○, HDPE-maleic anhydride; ▲, PP-acrylic acid; △, PP-maleic anhydride. (Ref. 25.)

Table 4.5 Grafting of Glycidyl Methacrylate onto PE or Polytetrafluoroethylene and Bonding Strength[a]

	Grafting (%)	Bonding strength (kg/cm^2)
Polyethylene	0	2
	13	13
Polytetrafluoroethylene	0	1.5
	31	14

Source: Ref. 25.
[a]Adhesion, epoxy resin/polyamide resin; curing condition, room temperature for 2 days.

Figure 4.15 Stabilization of grafted surface against accelerated weathering without water spray. An electron-induced graft of a carbon black-containing low-density PE sheet was stabilized with three different combinations of antioxidants and ultraviolet absorbers. ○, unstabilized, with water spray; □, unstabilized, without water spray; ●, ▲, ■, stabilized, without water spray. (Ref. 23.)

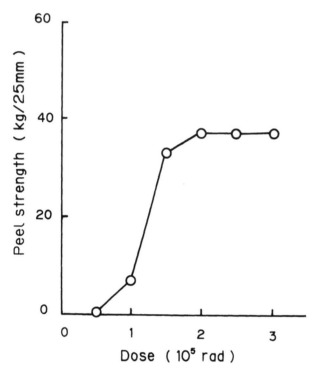

Figure 4.17   Peel strength versus dose of polytetrafluoroethylene
grafted with methyl acrylate.   (Ref. 26.)

4.4   PHOTOGRAFTING POLYMERIZATION

Methods for photografting polymerization are roughly divided into
liquid- and gas-phase methods, depending on the way of supplying
monomers.   They can also be classified by the manner of initiation:
(1) direct initiation by light, (2) initiation by a sensitizer, and (3)
initiation by backbone polymer containing photoactive groups.

In case 1 above, radicals are produced in the backbone polymer
with the aid of ultraviolet light alone.   Because of the low energy,
initiation does not occur efficiently in some polymers.   In case 2,
the sensitizer absorbs light and then gives energy to the backbone
polymer to produce radicals on the backbone; or the backbone polymer is attacked by radicals formed in the sensitizer itself or as a
result of its interaction with surrounding molecules.   In case 3,
active groups that absorb light are originally contained in the polymer substrate, and light works to activate them to provide radicals.

4.4.1 Application to Surface Modification and Effect on Adhesion

In general, grafting activity depends on the reactivity which in turn is based on the chemical structure of the substrate and the affinity between the substrate and the treatment solution. Some examples that demonstrate this are shown in Table 4.6 [27]. Typical processes of graft polymerization to a substrate by the method of case 1 include photografting of acrylamide to cellulose [28] and of acenaphthylene [29] or maleimide [30] to polyethylene films.

Photografting polymerization is often performed efficiently using a triplet sensitizer such as benzophenone. Photografting polymerization of styrene [31] and acrylamide [32] to polypropylene as well as that of acrylic acid and methyl methacrylate [33] to polypropylene and polyethylene have been conducted, demonstrating that they are introduced to the substrate surface at a high level of graft efficiency (see Table 4.6). It has been found, furthermore, the graft efficiency is increased sharply by the addition of water. Some relevant observations are shown in Figure 4.18 [34].

A process has been proposed to activate the material surface in advance of irradiation [35]. As shown in equation (1), a polyethylene film to which diethylamino groups have been added undergoes surface-graft polymerization when irradiated with ultraviolet light while in contact with a solution containing acrylic monomers and benzophenone.

$$PE \xrightarrow[Cl_2]{UV} \quad C\ell \xrightarrow{\overline{HN(CH_2 \cdot CH_2)_2}} \quad N(CH_2 \cdot CH_2)_2 \quad + BP$$

$$\xrightarrow{UV} \quad CH_2 \cdot CH_2 N \cdot \dot{C}H \cdot CH_2 \quad \longrightarrow \text{Graft polymerization} \tag{1}$$

This process takes advantage of the phenomenon that radicals which can initiate polymerization are produced quantitatively on the side of the amine when a tertiary amine and benzophenone coexist with the material under irradiation.

As described above, modified polymers are used effectively for fiberglass-reinforced plastics (FRP) in which fiberglass is incorporated. That is, nonpolar polymer such as polyolefine is poor in interfacial adhesion with fiberglass; for example, it is difficult for the reinforcing effect to be effective with a composite material made up of a combination of polypropylene and fiberglass treated with a normal silane coupling agent.

Pegoraro et al. reported [36] that acrylic acid graft-polymerized polypropylene and unmodified polypropylene were definitely different

Table 4.6 Summary of Solvent Effects on Surface Photografting of Polar Monomers onto Nonpolar Polymers

Solvent (S)	S-base polymer interactions	BP*3 + S radicals	Photografting	Surface concentration of graft chain	Surface modification effect	Proposed structure of grafted polymer
Polar	Weak	Yes	No	—	None	
	Weak	No	Yes, but slow	Low	Small	
Moderately polar	Moderate	Yes	No	—	None	
	Moderate	No	Yes, fast to medium	High	Large	
Nonpolar	Strong	Yes	Apparently yes (internal polymerization?)	Very low	Small	
	Strong	No	Yes, very fast	Low	Moderate	

Source: Ref. 27.

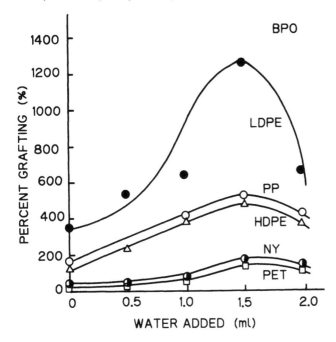

Figure 4.18   Effect of water on vapor-phase photografting of acrylic acid sensitized with BPO.   Irradiations were carried out at 60°C for 60 min.   (Ref. 34.)

from each other in terms of the interfacial adhesive strength (measured based on the conventional rules of mixture) of cut fiber-reinforced composite polymer. Sasaki et al. proposed [37] a method, which was useful as an industrial process, for effectively graft-polymerizing acrylic acid and maleic anhydride on polyolefine using an extruder commonly used for making pellets, with the results shown in Table 4.7. We see from these results that heat resistance and mechanical properties such as HDT (heat distortion temperature) and bending strength are greatly improved by graft-polymerizing a small amount of acid component on polypropylene.

Furthermore, as shown in Figure 4.19, in the case of unmodified polypropylene composite material, which is poor in adhesion, the unusual behavior is observed—a negative reinforcing effect as the creep deformation at 80°C increases with increased fiberglass content. In the case of polypropylene composite material modified with acid, which exhibits excellent interfacial adhesion, the composite generally shows only reduced deformation because of the effective action of

Table 4.7   Relation Between the Type of Monomer Added and Some Properties of FRPP

Type of monomer added	$[\eta]$ of PP[a]	HDT[b] (°C)	σ_F[b] (kg/mm^2)
None	1.29	130.0	8.7
Acrylic acid	1.27	156.0	11.8
Methacrylic acid	1.29	147.5	10.1
Crotonic acid	1.26	131.0	9.8
Maleic acid	1.18	151.5	9.5
Maleic anhydride	1.14	154.0	10.1
Citraconic acid	1.35	152.0	10.0
Citraconic anhydride	1.33	150.5	10.2
Itaconic acid	1.31	152.5	9.9
Itaconic anhydride	1.40	154.0	11.0
Methyl acrylate	1.16	132.5	10.0
Methyl methacrylate	1.26	139.5	9.7
Butyl acrylate	1.20	134.5	9.8
Vinyl-tris(methoxyethoxy) silane	—	133.0	9.0
γ-Glycidoxypropyltrimethoxy silane	—	152.0	10.3
γ-Aminopropyltriethoxy silane	—	144.5	9.8

Source: Ref. 37.

[a]PP, TA-2 modified with 0.001 mol % monomer and 0.15 wt % BPO.

[b]GF, 3E-401, 20%, dry-blend method.

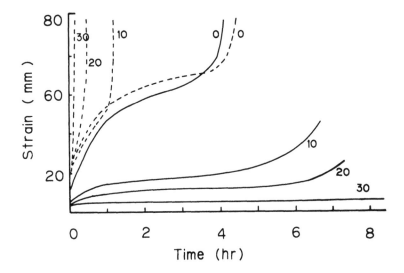

Figure 4.19   Creep properties of FRPP: —, unmodified PP; ---
PP grafted with 0.069% acrylic amide, numerals in figure show GF
content.   (Ref. 38.)

fiberglass as a stress absorber, and the usual decreasing deformation as the fiberglass content increases is observed.

These results exhibit not only the remarkable effect of graft
polymerization on creep performance but also the importance of interfacial adhesion between filler and matrix in the composite material.
Except for unsaturated carboxylic acid, it is known, for example,
that the affinity of polypropylene with fiberglass is improved by
graft polymerizing p-vinyl phenol in cooperation with treatment of
the glass by a silane compound; resistance to water, particularly, is
improved [39].

REFERENCES

1. S. Kaplan and A. Dilks, Characterization of Plasma-Polymerized
    Materials by Modern Spectroscopic Techniques, *J. Appl. Polym.
    Sci. Appl. Polym. Symp.*, *38*, 105 (1984).

2. A. Moshonov and Y. Avny, The Use of Acetylene Glow Discharge
    for Improving Adhesive Bonding of Polymeric Films, *J. Appl.
    Polym. Sci.*, *25*, 771 (1980).

3. H. Yasuda, T. S. Hsu, E. S. Brandt, and C. N. Reilley, Some
    Aspects of Plasma Polymerization of Fluorine-Containing Organic

Compounds; II. Comparison of Ethylene and Tetrafluoroethylene, *J. Polym. Sci. Polym. Chem. Ed.*, *16*, 415 (1978).

4. N. Inagaki and H. Yasuda, Adhesion of Glow Discharge Polymers to Metals and Polymers, *J. Appl. Polym. Sci.*, *26*, 3333 (1981).

5. M. Suzuki, A. Kishida, H. Iwata, and Y. Ikada, Graft Copolymerization of Acrylamide onto a Polyethylene Surface Pretreated with a Glow Discharge, *Macromolecules*, *19*, 1804 (1986).

6. H. K. Yasuda, A. K. Sharma, E. B. Hale, and W. J. James, Atomic Interfacial Mixing to Create Water Insensitive Adhesion, *J. Adhes.*, *13*, 269 (1982).

7a. PDA Engineering Co., Solve Composites Adhesion Problem, *Adv. Manuf. Technol.*, *9*, 7 (1987).

7b. H. Iwata, A. Kishida, M. Suzuki, Y. Hata, and Y. Ikada, Oxidation of polyethylene surface by corona discharge and subsequent graft polymerization, *J. Polym. Sci.*, *A*, Polym. Chem., *26*. 3309 (1988).

8. L. A. Wall Factors Influencing the Behavior of Polymers Exposed to High-Energy Radiation, *J. Polym. Sci.*, *Lett. Ed.*, *17*, 141 (1955).

9. S. L. Madorsky and S. Straus, Thermal Degradation of Polymers as a Function of Molecular Structure, *J. Res. Natl. Bur. Stand.*, *53*, 361 (1954).

10. A. Chapio, The Action of Gamma Rays on Polymers in the Solid State; I. Reticulation of Polyethylene, *J. Chim. Phys.*, *52*, 246 (1955).

11. P. Y. Feng and J. W. Kennedy, Electrical and Chemical Effects of β-Radiation in Polystyrene, *J. Am. Chem. Soc.*, *77*, 847 (1955).

12. A. Chapiro, The Action of Gamma Rays on Polymers in the Solid State; III. Irradiation of Polyvinylchloride, *J. Chim. Phys.*, *53*, 895 (1956).

13. L. A. Wall and R. E. Folorin, Polytetrafluorethylene—A Radiation-Resistant Polymer, *J. Appl. Polym. Sci.*, *Lett. Ed.*, *2*, 251 (1959).

14. M. Dole, C. D. Keeling, and D. G. Rose, The Pile Irradiation of Polyethylene, *J. Am. Chem., Soc.*, *76*, 4304 (1954).

15a. Y. Ikada, M. Suzuki, M. Taniguchi, H. Iwata, W. Taki, H. Miyake, Y. Yonekawa, and H. Handa, Interaction of Blood with Radiation-Grafted Materials, *Radiat. Phys. Chem.*, *18*, 1207 (1981).

15b. M. Suzuki, Y. Tamada, H. Iwata, and Y. Ikada, Polymer Sur-
face Modification to Attain Blood Compatibility of Hydrophobic
Polymers, *Phys.- Chem. Aspects of Polym. Surfaces, Vol. 2,*
Ed. by K. L. Mittal, Plenum Pub. (1983).

16. V. A. Postnikov, N. Ju. Lukin, B. V. Maslov, and N. A.
Plate, The Simple Preparative Synthesis of Graft Copolymers
of Polyethylene-Acrylamide, *Polym. Bull.,* 3, 75 (1980).

17. J. K. Rieke, G. M. Hart, and F. L. Saunders, Graft Copolymer of Polyethylene and Acrylic Acid, *J. Polym. Sci., C-4,*
589 (1963).

18. S. Yamakawa, Surface Modification of PE by Radiation—Induced
Grafting for Adhesive Bonding: 1. Relation Between Adhesive
Bond Strength and Surface Composition, *J. Appl. Polym. Sci.,*
20, 3057 (1976).

19. F. Yamamoto and S. Yamakawa, Surface Grafting of Polyethylene by Mutual Irradiation in Methyl Acrylate Vapor: I.
γ-Irradiation, *J. Polym. Sci. Polym. Chem. Ed.,* 16, 1883
(1978).

20. F. Yamamoto, S. Yamakawa, and Y. Kato, Surface Grafting of
Polyethylene by Mutual Irradiation in Methyl Acrylate Vapor:
II. High-Energy Electrons Irradiation, *J. Polym. Sci. Polym.
Chem. Ed.,* 16, 1897 (1978).

21. S. Yamakawa and F. Yamamoto, Surface Modification of Polyethylene by Radiation Induced Grafting for Adhesive Bonding:
IV. Improvements in Wet Peel Strength, V. Comparison with
Other Surface Treatment, *J. Appl. Polym. Sci.,* 25, 25 41
(1980).

22. S. Yamakawa, Surface Modification of Polyethylene by Radiation
Induced Grafting for Adhesive Bonding, *Macromolecules,* 9, 754
(1976).

23. S. Yamakawa and F. Yamamoto, Surface Modification of Poly-
ethylene by Radiation Induced Grafting for Adhesive Bonding:
III. Oxidative Degradation and Stabilization of Grafted Layer,
J. Appl. Polym. Sci., 22, 2459 (1978).

24. K. Manaka and T. Tomioka, Improvement of Adhesive Properties of Plastics by Radiation Graft Polymerization, *J. Appl.
Polym. Sci.,* 9, 3635 (1965).

25. F. Ide, Adhesion and Graft Polymer, *Polym. Appl. Jpn.,* 28,
327 (1978).

26. S. Yamakawa, Surface Modification of Fluorocarbon Polymers by
Radiation-Induced Grafting for Adhesive Bonding, *Macromolecules,*
12, 1222 (1979).

27. S. Tazuke, Thin-Layer Photografting as a Method of Polymer Surface Modification, *Poly. Plast. Technol. Eng.*, *14*, 107 (1980).

28. A. H. Reine and J. C. Arthur, Jr., Photoinitiated Polymerization of Methacrylamide with Cotton Cellulose, *Text. Res. J.*, *42*, 155 (1972).

29. K. Hayakawa, K. Kawase, and H. Yamakita, Reaction of Acenaphthylene Vapor in the Presence of Polymer Films Under γ- or Ultraviolet Irradiation, *J. Polym. Sci. A-1*, *10*, 2463 (1972).

30. K. Hayakawa, K. Kawase, and H. Yamakita, Further Studies on the Ultraviolet Graft Copolymerization of Maleimide by Vapor-Phase Method, *J. Polym. Sci. Polym. Chem. Ed.*, *12*, 2603 (1974).

31. C. H. Ang, N. P. Davis, L. Garnett, and N. T. Yen, The Nature of Bonding During Film Formation on Polyolefins and Cellulose Using UV and Ionizing Radiation Initiation, *Radiat. Phys. Chem.*, *9*, 831 (1977).

32. S. Tazuke and H. Kimura, Graft Polymerization of Hydrophilic Monomers onto Various Polymer Films, *J. Polym. Sci. Polym. Lett. Ed.*, *16*, 497 (1978).

33. Y. Ogiwara, M. Kanda, M. Takumi, and H. Kubota, Photosensitized Grafting on Polyolefin Films in Vapor and Liquid Phases, *J. Polym. Sci. Polym. Lett. Ed.*, *19*, 457 (1981).

34. Y. Ogiwara, K. Torikoshi, and H. Kubota, Vapro Phase Photografting of Acrylic Acid on Polymer Film; Effects of Solvent Mixed and Monomer, *J. Polym. Sci. Polym. Lett. Ed.*, *20*, 17 (1982).

35. J. F. Kinstle and S. L. Watson, Jr., Photoassisted Modification of and Grafting to Polyethylene, *Polym. Sci. Technol.*, *10*, 461 (1977).

36. M. Pegoraro, G. Pagani, P. Clerici, and A. Penati, Interaction between Short Subcritical Length Fibers and Polymer Matrix, *Fibre Sci. Technol.*, *10*, 263 (1977).

37. I. Sasaki and F. Ide, Mechanism of Reinforcement with Glass-fiber in Polypropylene Grafted with Unsaturated Carboxylic Acid, *Kobunshi Ronbunshu*, *38*, 75 (1981).

38. I. Sasaki and F. Ide, Effect of Grafting of Unsaturated Carboxylic Acid on Glass Fiber-Reinforced Polypropylene, *Kobunshi Ronbunshu*, *38*, 67 (1981).

39. S. Yamamoto, S. Onishi, and A. Takahashi, *Studies on Adhesive Properties of Poly(p-Vinylphenol)*, No. 128, Textile and Polymer Material Institute of Japanese Government, 1981, p. 83.

5

Modification of Inorganic Fillers for Composite Materials

5.1 INTRODUCTION

Generally, there are large differences between the matrix and filler constituting a composite in such properties as density, elastic modulus, thermal expansion, and surface energy. Desirable properties cannot be obtained merely by more mixing and dispersion of raw material; complete adhesion at the interface is also essential. It is therefore necessary, to improve the surface properties properly in order to increase the affinity between the filler and the matrix to allow them to form a well-integrated material. In this chapter we deal with improvement methods and adhesion properties of inorganic filler surfaces. In general, an inorganic filler tends to have (1) a large specific surface, and (2) as a result, a large surface energy, and (3) a complicated surface structure with structural irregularities. All these properties indicate that the surface of the inorganic filler is active both physically and chemically. The surface characteristics of an inorganic filler depend on the general properties of the solid surface as well as on the intrinsic properties (such as composition and morphology) of the filler.

Most of the inorganic fillers in common use in industrial manufacturing have hydrophilic surfaces. Efforts have been under way for many years ago to convert hydrophilic surfaces into hydrophobic and lipophilic surfaces on the basis of their surface activity, to improve the wetting properties and dispersibility of fillers on their matrices. Most recent studies focus on enhancing the adhesion between fillers and matrices by increasing the functionality of the filler surface.

5.2 SURFACE IMPROVEMENTS BY COUPLING AGENTS

5.2.1 Types and Properties of Coupling Agents

The structure of a silane coupling agent is expressed by the general formula $(RO)_3-Si-R'$, where the (RO) group represents functional groups hydrolyzed to give a silanol group (e.g., methoxy, ethoxy), and R' is for those groups that have affinity for and display reactivity to the matrix materials (i.e., nonhydrolizable materials, including vinyl, epoxy, and amino groups). The reactivity to thermosetting resins such as unsaturated polyester and epoxy resins is taken into account in selecting the most suitable group for R'. Table 5.1 lists the structural formulas and properties of major commercial silane coupling agents [1].

The structure of a titanate coupling agent is expressed by the general formula $RO-Ti-(OXRY)_3$. Various types with different hydrophilic (RO) groups and lipophilic $(OXRY)$ groups are available commercially. The structural formulas and properties of major titanate coupling agents are listed in Table 5.2 [2]. These agents are categorized into four groups, depending on the type of hydrophilic group, as shown in Table 5.3. The monoalkoxy type (1), No. 100 chelate type (2), and No. 200 chelate type (3) contain an isopropoxy group, oxyacetic acid residue, and ethylene glycol residue, respectively, while the coordinate type (4) is produced by adding phosphites to conventional tetraalkoxytitanates. KR-38R, KR-138S, and KR-238S, for example, have the same lipophilic group (i.e., dioctyl pyrophosphate), but with different hydrophilic groups. The conventional alkoxy-type agents can be used only in lipophilic conditions; in a chelate-type agent, a hydrophilic group is substituted to form an addition salt with an amine to enable it to be used in water-soluble conditions. Coordinate-type agents have increased resistance to hydrolysis as well as other favorable properties, such as the high fire resistance characteristic of most phosphorus compounds.

Regarding the prominent features of these agents, it is widely known that silane coupling agents act to increase strength and rigidity, while titanate coupling agents are very useful for improving flexibility and processability. The former are particularly effective for fillers composed of silicon; the latter can be applied to a wide variety of fillers. The cause of this difference has not been clarified, offering an interesting subject for future studies. However, the wide differences between their electronic states seems to provide a clue. Si consists of $1s^2$, $2s^2$, $2p^6$, $3s^2$, and $2p^2$, and Ti has $1s^2$, $2s^2$, $2p^6$, $3s^2$, $3p^6$, $3d^2$, and $4s^2$, showing that it is a typical transition metal. With the same valence of 4, Si persists in a state with a coordination number of 4 and hardly allow its third orbit to be

Table 5.1 Structures and Physical Properties of Silane Coupling Agents

Chemical name	Structure	Molecular weight	Specific gravity at 25°C
Vinyltrichlorosilane	$CH_2=CHSiCl_3$	161.5	1.26
Vinyltriethoxysilane	$CH_2=CHSi(OC_2H_5)_3$	190.3	0.93
Vinyltri(β-methoxyethoxy)-silane	$CH_2=CHSi(OC_2H_4OCH_3)_3$	280.4	1.04
γ-Glycidoxypropyltrimethoxy-silane	$CH_2\text{—}CHCH_2OCH_2CH_2CH_2Si(OCH_3)_3$ (with O bridging)	236.1	1.07
γ-Methacryloxypropyltrimethoxy-silane	$CH_2=\underset{\underset{}{CH_3}}{C}\text{—}\underset{\underset{O}{\parallel}}{C}\text{—}O\text{—}C_3H_6Si(OCH_3)_3$	248.1	1.04
N-(β-Aminoethyl)γ-aminopropyl-trimethoxysilane	$H_2NC_2H_4NHC_3H_6Si(OCH_3)_3$	222.1	1.03
N-(β-Aminoethyl)γ-aminopropyl-methyldimethoxysilane	$H_2NC_2H_4NHC_3H_6Si(OCH_3)_2$ CH_3	206.1	0.98
γ-Chloropropyltrimethoxysilane	$ClC_3H_6Si(OCH_3)_3$	198.5	1.08
γ-Mercaptopropyltrimethoxysilane	$HSC_3H_6Si(OCH_3)_3$	196.1	1.06
γ-Aminopropyltriethoxysilane	$H_2NC_3H_6Si(OC_2H_5)_3$	221.0	0.94

Source: Ref. 1.

Table 5.2 Structures and Physical Properties of Titanate Coupling Agents

Chemical name	Structure	Molecular weight	Specific gravity (25°C)
Isopropyltriisostearoyl-titanate TTS (2-3099)	$\left[\text{CH}_3-\text{CH}-\text{O}-\text{Ti}\right]$ with CH$_3$ branch, $-\text{O}-\overset{\overset{\text{O}}{\|}}{\text{C}}-\text{C}_{17}\text{H}_{35}$, subscript 3	957	0.94
Isopropyltridodecylbenzene-surfonil titanate 9S (2-3252)	$\left[\text{CH}_3-\text{CH}-\text{O}-\text{Ti}\right]$ with CH$_3$ branch, $-\text{O}-\overset{\overset{\text{O}}{\|}}{\underset{\underset{\text{O}}{\|}}{\text{S}}}-\text{C}_6\text{H}_4-\text{C}_{12}\text{H}_{25}$, subscript 3	1083	1.09
Isopropyl-tri(dioctylpyro-phosphate)titanate 38S	$\left[\text{CH}_3-\text{CH}-\text{O}-\text{Ti}\right]$ with CH$_3$ branch, $-\text{O}-\overset{\overset{\text{O}}{\|}}{\text{P}}-\text{O}-\overset{\overset{\text{O}}{\|}}{\underset{\underset{\text{OH}}{}}{\text{P}}}-(\text{O}-\text{C}_8\text{H}_{17})_2$, subscript 3	1311	1.10
Tetraisopropyl-bis(dioctyl-phosphite) titanate 41B	$\left[\text{CH}_3-\text{CH}-\text{O}-\right]_4 \text{Ti}\cdot[\text{P}(\text{O}-\text{C}_8\text{H}_{17})_2\text{OH}]_2$ with CH$_3$ branch	897	0.94

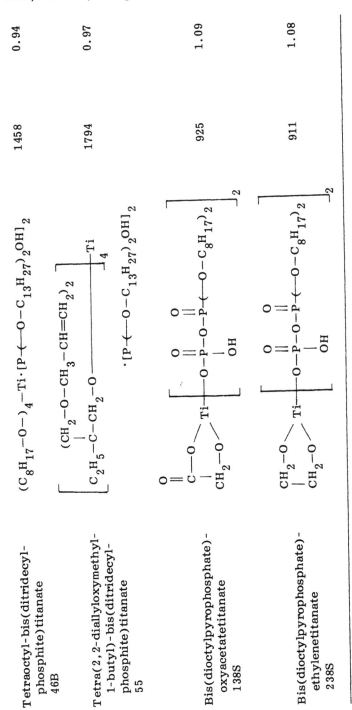

Tetraoctyl-bis(ditridecyl-phosphite)titanate 46B	$(C_8H_{17}-O-)_4-Ti\cdot[P-(-O-C_{13}H_{27})_2OH]_2$	1458	0.94
Tetra(2,2-diallyloxymethyl-1-butyl)-bis(ditridecyl-phosphite)titanate 55	$[C_2H_5-C-CH_2-O-\ (CH_2-O-CH_3-CH=CH_2)_2\ -Ti]_4\cdot[P-(-O-C_{13}H_{27})_2OH]_2$	1794	0.97
Bis(dioctylpyrophosphate)-oxyacetatetitanate 138S		925	1.09
Bis(dioctylpyrophosphate)-ethylenetitanate 238S		911	1.08

Source: Ref. 2b.

Table 5.3 Generalized and Typical Chemical Structures of Four Classes of Titanates

Class	Molecular structure
Monoalkoxy	$i-\text{PrO}-\text{Ti}\!-\!(\text{OR})_3$
Chelate 100 series	$$\begin{array}{c} \text{O} \\ \| \\ \text{C}-\text{O} \\ | \\ \text{H}_2\text{C}-\text{O} \end{array} \!\!\! \rangle \text{Ti}\!-\!(\text{OR})_2$$
200 series	$$\begin{array}{c} \text{H}_2\text{C}-\text{O} \\ | \\ \text{H}_2\text{C}-\text{O} \end{array} \!\!\! \rangle \text{Ti}\!-\!(\text{OR})_2$$
Coordinate	$(i\text{-PrO})_4-\text{Ti}\cdot[\text{P(OR)}_2\text{OH}]_2$

involved, whereas Ti can even assume a loose structure with a co-ordination number of 6 and permit various ligands to coordinate with its third orbit. Many inorganic fillers are likely to act as ligands, and it is therefore inferred that the difference between the properties of Si and Ti may arise from the variation in their ability to receive ligands, although no evidence of this has yet been obtained.

5.2.2 Action Mechanism and Effect of Coupling Agents

Some ideas have been proposed for the mechanism of action of silane coupling agents on the surfaces of inorganic substances the most important of which are:

1. The silane coupling agent undergoes a chemical reaction with the surface of the inorganic substance to form an SiOM bond (M: Si atom in glass, silica, etc., or metallic atom) [3].
2. The Silane coupling agent is physically adsorbed on the inorganic surface [4].
3. The Si—OH group on the glass surface forms a hydrogen bond with the silanol group derived from the silane coupling agent [5].

4. The silane coupling agent forms a sheathlike structure around the glass fiber [6].
5. A reversible equilibrium reaction takes place between the hydroxyl group on the inorganic surface and the silanol group derived from the silane coupling agent [7].

Despite the various studies, the mechanism of action of these agents has not yet been fully clarified, suggesting that the interface conditions within a composite are highly complicated. This complexity probably results from the involvement of various factors, such as the methods and conditions for processing the silane coupling agent as well as the type and surface conditions of the inorganic substance. Recently, the structure and thickness of the adsorption layer of silane have been determined using improved instrumental analysis methods [8–11]. Another factor adding to the complexity is the structural change in silane coupling agents when in aqueous solutions. The alkoxyl group in silane coupling agents and the silanol group resulting from their hydrolysis are essentially unstable in aqueous solutions, leading to a structural change over time. Analysis should also be carried out to clarify the relations of the structural changes to the properties of the functional groups in silane coupling agents, the pH values of their aqueous solutions, and the silane concentrations. This is one of the most important questions remaining to be solved [12].

A brief discussion follows on the chemical bonding theory described in (1) above, which is more widely known than are the other theories. As shown in Figure 5.1, the hydrolyzable group in

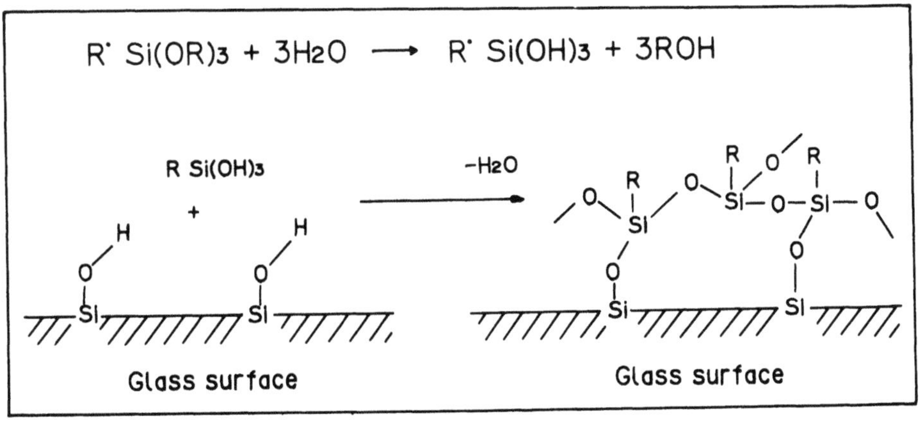

Figure 5.1   Chemical bonding theory.   (Ref 2a.)

Table 5.4   Relationship Between SP Value and Flexural Strength

	SP value $(\text{cal}/\text{cm}^3)^{1/2}$	Flexural strength (kg/mm^2)
Vinyl	6.5	131.5
Ethyl	7.0	73.1
Dodecyl	8.0	69.6
Mercaptopropyl	8.6	158.2
Aminopropyl	8.7	96.3
Methacryloxypropyl	9.0	201.0
Glycidoxypropyl	9.3	179.3
Phenyl	9.2	75.9
Cyanopropyl	10.5	30.9

Source: Ref. 13.

a silane coupling group is first hydrolyzed to give a silanol group, which then undergoes a condensation reaction with the silanol group on the surface of the inorganic material (i.e., glass, silica, etc.) to form covalent bonding between the coupling agent and the material. On the other hand, the nonhydrolyzable functional group in the coupling agent bonds to the matrix through a chemical reaction. Thus the coupling agent serves to increase the strength of the composite by providing chemical bonding to connect the glass surface and the matrix.   As an example, Table 5.4 shows relations of the solubility parameter (SP value) of silane coupling agents with the strength of unsaturated polyester–glass fiber composites [13]. There is no correlation between the strength of the composites and the difference in the SP value of the unsaturated polyesters and the silane coupling agents, whereas methacryoxypropyl silane and vinyl silane, which have unsaturated groups, act to increase the strength.   This indicates that in this composite system, the strength depends on the reactivity of the agents to the unsaturated polyesters rather than on the wetting effects of the silane treatment, strongly suggesting the formation of chemical bonding between the resins and the coupling agents.   Thus the interfacial chemical bonding theory is generally accepted for composite materials consisting of thermo- setting resin matrices [14].   Mechanisms such as the following have also been proposed: (1) wetting on the matrix polymer is enhanced [15,16]; (2) compatibility with the matrix polymer is improved; and

Table 5.5   Effect of Silane Coupling Agents on Adhesion Between Al and HDPE

Silane coupling agents	180° peel strength (kg/cm)	
	press (200°C × 10 min)	press (150°C × 10 min)
$(CH_3O)_3Si(CH_2)_3NHCH_2CH_2NH_2$	3.4	1.8
$(CH_3OC_2H_4O)_3SiCH=CH_2$	4.1	—
$(CH_3O)_3Si(CH_2)_3OCH_2\overset{O}{\overbrace{CHCH_2}}$	3.4	0.6
$(CH_3O)_3Si(CH_2)_3OCOC(CH_3)=CH_2$	5.0	2.0
$(t\text{-}C_4H_9O)_3SiCH=CH_2$	7.3	0.6
$(CH_3O)_3Si(CH_2)_3NHCH_2CH=CH_2$	7.4	5.2
$(C_2H_5O)_3Si(CH_2)_3NHCH_2CH=CH_2:\ HCl$	6.9	5.6
$(CH_3O)_3Si(CH_2)_3NHCH_2CH_2NHCH_2CH=CH_2$	6.9	5.7
$(CH_3O)_3Si(CH_2)_3NHCH_2CH_2NHCH_2COOH$	4.4	3.1
$(C_2H_5O)_3Si(CH_2)_3NHCOCH=CHCOOH$	5.8	3.6
$(CH_3O)_3Si(CH_2)_3SCH_2CH_2COOH$	6.8	5.9

Source: Ref. 2a.

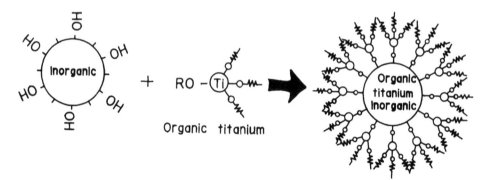

Figure 5.2   Coupling mechanism of titanium coupling agent. (Ref. 2a,23)

(3) friction between the matrix polymer and the glass fiber is increased [17].

It has been reported that compared to thermosetting resins, silane coupling agents have less effect on thermoplastic resins, polyolefins in particular, for which there are no specific reactions that may involve a silane coupling agent.   Active studies are being carried out

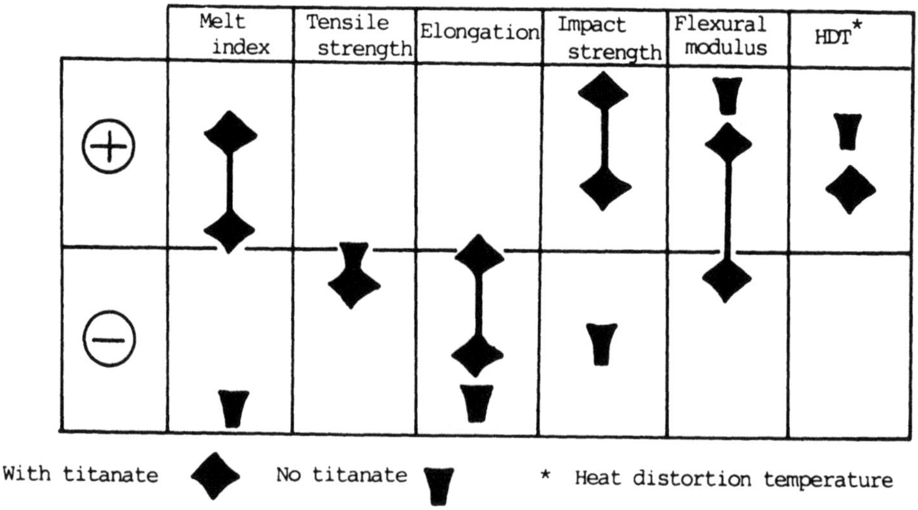

	Melt index	Tensile strength	Elongation	Impact strength	Flexural modulus	HDT*
⊕						
⊖						

With titanate ◆   No titanate ▼   * Heat distortion temperature

Figure 5.3   Generalized effects of high filler loading on various properties of polyolefin.   (Ref. 2a.)

Table 5.6 Results of Contact Angles Measured by Tablet and Penetration Rate Methods

Dispersion medium / Measurement / Sample	Contact angle(°)(23°C)					
	Water		Toluene		Liquid paraffin	
	Tablet method	Penetration rate method	Tablet method	Penetration rate method	Tablet method	Penetration rate method
Calcium carbonate (Untreated)	20	20 > (Occurrence of a crack)	14	48	30	56
Calcium carbonate (Treated with TC-1)	110	90 < (No penetration)	25	20 > (Occurrence of a crack)	22	20 > (Occurrence of a crack)
Calcium carbonate (Treated with TC-2)	26	87	14	20 > (Occurrence of a crack)	24	20 > (Occurrence of a crack)
Calcium carbonate (Treated with TC-3)	22	87	20	40	28	51
Silica (Untreated)	21	73	16	0	34	35
Silica (Treated with TC-1)	110	90 < (No penetration)	14	20 > (Occurrence of a crack)	25	20 > (Occurrence of a crack)
Silica (Treated with TC-2)	117	90 < (No penetration)	13	20 > (Occurrence of a crack)	32	20 > (Occurrence of a crack)
Silica (Treated with TC-3)	25	79	14	63	38	77
Talc (Untreated)	64	79	14	56	26	51
Talc (Treated with TC-1)	89	90 < (No penetration)	15	35	22	32
Talc (Treated with TC-2)	68	78	16	37	21	40
Talc (Treated with TC-3)	56	70	17	54	20	46

Source: Ref. 25.

Table 5.7 Degrees of Dispersion Observed Under the Microscope and by Sedimented Powder Volume

Dispersion Medium / Observation	Dispersion (25°C)					
	Water		Toluene		Liquid paraffin	
	Microscope	Sedimentation Volume	Microscope	Sedimentation volume	Microscope	Sedimentation volume
Calcium carbonate (Untreated)	◎	◎	X	X	X	X
Calcium carbonate (Treated with TC-1)	X	X	◎	◎	◎	◎
Calcium carbonate (Treated with TC-2)	△	△	○	○	○	○
Calcium carbonate (Treated with TC-3)	○	△	△	△	○	○
Silica (Untreated)	◎	◎	X	X	X	X
Silica (Treated with TC-1)	X	X	◎	◎	◎	◎
Silica (Treated with TC-2)	△	X	○	○	○	○
Silica (Treated with TC-3)	△	○	△	△	○	△
Talc (Untreated)	○	○	○	△	○	○
Talc (Treated with TC-1)	△	X	○	○	○	○
Talc (Treated with TC-2)	○	○	○	△	○	○
Talc (Treated with TC-3)	○	○	○	○	△	○

◎ : Well   ○ : slightly well   △ : slightly bad   X : Bad

Source: Ref. 25.

in this field, bringing about the development of an amino silane with an unsaturated bond, carboxylic acid functional silane, cationic silane [18,19], silyl peroxide [20,21], and aminimide [22]. Table 5.5 shows the effects of these silane coupling agents on the properties of high-density polyethylene-aluminum sandwich panels.

Many studies have also been conducted on the action mechanism of titanate coupling agents, which were developed relatively recently, although sufficient evidence has not been obtained for the bonding mechanisms or interactions between titanate coupling agents and inorganic fillers or matrix resins. Bonding between the agents and inorganic fillers is generally interpreted based on a chemical bonding theory as shown in Figure 5.2, although it has not been demonstrated appropriately [23]. There are, however, many reports that propose indirect evidence which suggests that the agents do interact somewhat with inorganic fillers [24]. Few titanate coupling agents have nonhydrolyzable groups that can undergo a chemical reaction with the matrix resins. This indicates that for the most part, interactions of titanate coupling agents with matrix resins result from the entanglement of molecules of the former with relatively long-chain hydrocarbon radicals of the latter and that the entanglement is responsible for the surface modification effect. Figure 5.3 shows typical properties of a composite composed of a polyolefin resin and an inorganic filler treated with a titanate coupling agent. Results on the wetting and dispersion properties are given in Tables 5.6 and 5.7, respectively. It can be seen that silane coupling agents are effective for increasing the rigidity, while titanate coupling agents act to improve the processability and flexibility.

5.3 FUNCTIONAL GROUPS ON INORGANIC FILLER SURFACE

The functional groups on the surface of an inorganic filler are an important factor in investigating interactions between the filler and a matrix polymer at the interface because they correspond to the end groups of the polymer. Such groups have not been identified in all inorganic fillers. Functional groups on the carbon black surface have been studied for many years ago by organic chemical methods, showing that they exist only at the end of the graphite crystallite layer constituting the carbon black material.

There are three types of oxygen-containing functional groups: acid, neutral, and basic. Carbon generally forms acid surface oxides when heated in an oxygen atmosphere, and it forms basic surface oxides when heated in an inert gas. Both acid and basic surface oxides exist in any carbon material, although the latter always in smaller amounts. The acid functional groups include carboxyl and phenolic hydroxyl groups, while the neutral ones include carbonyl and quinone groups. However, the structures of the basic functional groups are not known in detail. Table 5.8 shows the distributions of oxygen-containing surface functional groups that exist on the carbon black surface [25,26].

It is known that two types of groups, siloxane and silanol, exist on the surface of silicic acid. Silanol groups are left in large amounts within the filler particles formed through a condensation reaction of low-molecular-weight silicic acids. The following surface functional groups will be produced, depending on the crystal structure formed [27]:

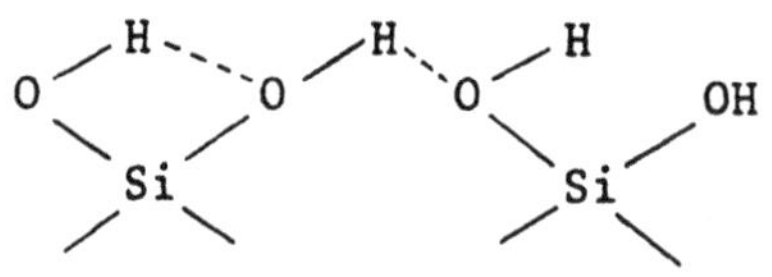

Cristobalite 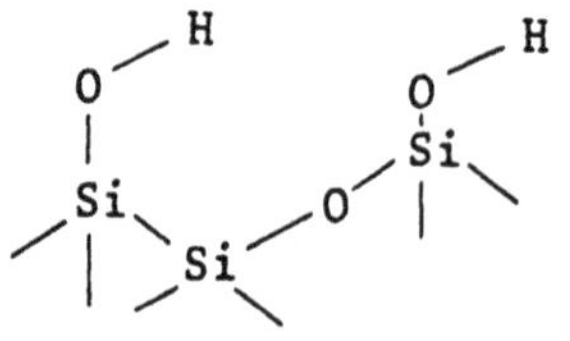   Tridymite

It is accepted that the cristobalite and tridymite typed coexist on the amorphous silicic acid surface in a ratio of about 60% to 40%. Table 5.9 lists measurements made by various techniques for the silanol group in dry white carbon (Aerosil) [28]. The existence of hydroxyl groups has been observed on the surface of rutile- and anatase-type titanium dioxides by the infrared-absorbing analysis method [29,30]. The difference between the adsorbed H_2O on a rutile surface and those on an anatase surface can be determined by nuclear magnetic resonance absorption at 300°K [31]. The following models have been proposed for the structure of the titanium dioxide surface under oxidation and reduction conditions [32]:

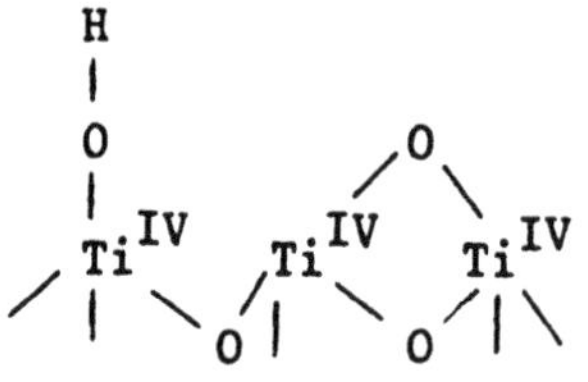

surface in   surface in
oxidation condition   reduction condition

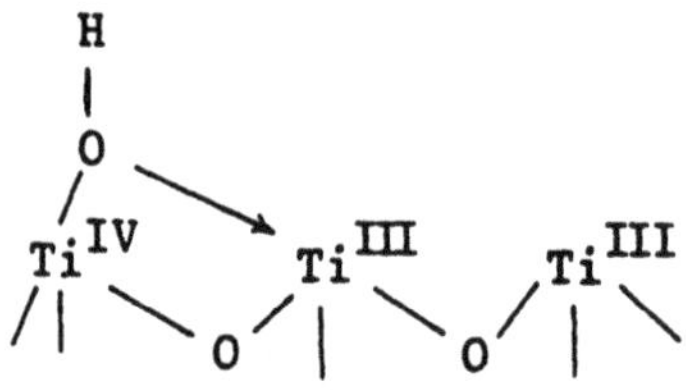

Although hydroxyl groups also exist on an alumina surface, alumina and its hydrates assume a great number of transformed structures,

Table 5.8   Surface Group Distributions

Sample	Hydrogen >—H		Hydroxyl >—OH		Quinone >=O		Carboxylic >—CO_2H		Lactone >—CO_2	
	meq/g	%	meq/g	%	meq/g	%	meq/g	%	meq/g	%
FF-C	3.94	90	0.30	7	0.03	1	0.04	1	0.06	1
FF-A-3	3.69	83	0.40	9	0.13	3	0.11	2	0.13	3
FF-B-3	2.92	75	0.67	17	0.02	1	0.13	3	0.16	4
HAF-C	1.66	78	0.27	13	0.11	5	0.01	0	0.09	4
HAF-A-2	0.71	41	0.53	31	0.22	13	0.12	7	0.13	8
HAF-B-2	0.59	39	0.61	41	0.00	0	0.12	8	0.18	12
EPC	4.65	73	0.86	13	0.64	10	0.05	1	0.20	3

Source: Ref. 26b.

Table 5.9   Density and Silanol Group of Aerosil Surface

Reaction	Specific surface (m^2/g)	Silanol groups	
		(meq/100 g)	Packing density $(OH/100 \text{ Å}^2)$
Reaction at 1000°C (free H_2O, K-Fischer titration)	180	103	3.45
Reaction with $SOCl_2$	178	62	2.10
Zerewitinoff method by CH_3MgI or CH_3Li	178	122	4.12
Reaction with $SOCl_2$	165	87	3.17
Titration by NaOH	145	57	2.38
Reaction with BCl_3	145	56	2.34
Al^{3+} absorption from $Al(OH)_2Cl(aq)$	145	60	2.50
UO^{2+} absorption from $UO_2(CH_3CO_2)_2$ (pH 5.4)	145	52	2.17
Methyration of CH_3OH at 200–250°C	145	55	2.29
IR spectroscopic	145	115	4.77

Source: Ref. 28.

resulting in different surface properties [33].   These infrared spectroscopic observations suggest that hydroxyl groups exist in all inorganic
fillers and that their surfaces, which are hydrophilic in varying degrees, must not be highly wettable by matrix polymers.   Studies have
been carried out on reactions involving the surface hydroxyl group or
other oxygen-containing groups to determine their use for surface improvement.

   The following exchange reaction takes place when silicic acid is immersed in an aqueous solution containing transition metal ions:

$$M^{n\cdot} + mSi(OH) \rightleftharpoons M(SiO)_m^{(n-m)\cdot} + mH\cdot$$

This exchange reaction proceeds less rapidly in the order Zn(II) >
Cu(II) > Ni(II) > Co(II) > Mn(II).   Except for Zn, a metal ion with

a larger diameter and larger density of electric charge tends to undergo the ion-exchange reaction more rapidly [34]. Calcium hydroxide also reacts rapidly, as shown below.

$$2Si(OH) + Ca(OH)_2 \rightleftharpoons (SiO)_2Ca + 2H_2O$$

This reaction is actually used for surface modification. It is also possible to make a silicic acid surface undergo the following reaction in a basic solution of aluminum chloride:

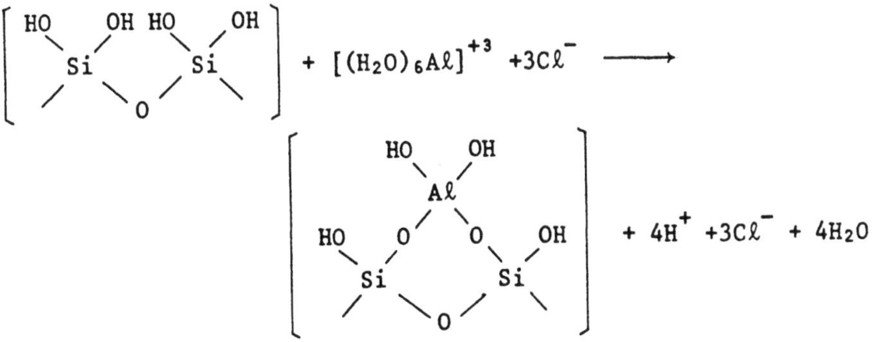

Various chemical reactions have been attempted in order to identify functional groups on the surfaces of inorganic fillers. The following, for example, are important reactions for silanol group contained in silicic acid or silicates.

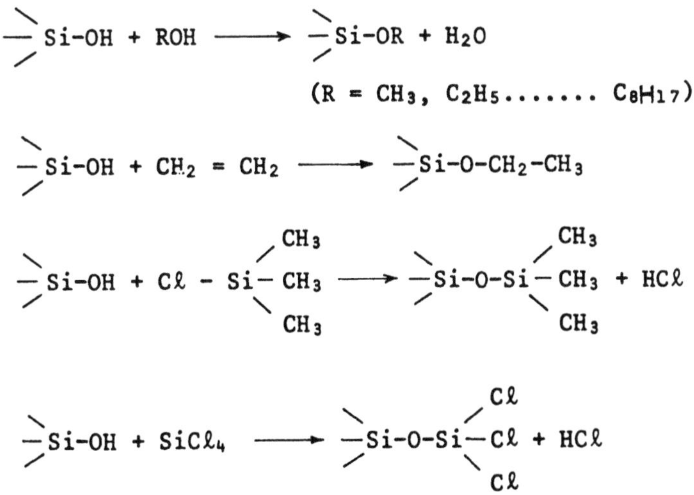

$$\ce{\backslash Si-OH + ROH -> \backslash Si-OR + H_2O}$$

$$(R = CH_3, C_2H_5 \ldots\ldots C8H_{17})$$

$$\ce{\backslash Si-OH + CH_2 = CH_2 -> \backslash Si-O-CH_2-CH_3}$$

$$\ce{\backslash Si-OH + Cl-Si(CH_3)_3 -> \backslash Si-O-Si(CH_3)_3 + HCl}$$

$$\ce{\backslash Si-OH + SiCl_4 -> \backslash Si-O-Si(Cl)_3 + HCl}$$

頁

Chapter 5

Table 5.10   Effect of Lipophilic Treatment on Various Metal Oxide Powders[a]

	Water	Benzene	Benzene/water	Butanol/water	Water/CCl$_4$
Silica gel	o	o	x/o	x/o	o/x
C$_2$H$_5$OH-treated silica gel	x	o	o/x	o/x	x/o
C$_2$H$_5$OH-treated silica gel	x	o	o/x	o/x	x/o
Magnetite	o	o	Δ/o	Δ/o	o/x
C$_2$H$_5$OH-treated magnetite	x	o	o/x	o/x	x/o
ZnO	o	o	o/x	o/x	x/Δ
C$_2$H$_5$OH-treated ZnO	o	o	o/x	x/⊖	x/o
Al$_2$O$_3$	o	o	o/o	o/o	x/o
C$_2$H$_5$OH-treated Al$_2$O$_3$	o	o	o/x	⊖/x	x/o
CCl$_4$-treated Al$_2$O$_3$	x	o	o/x	o/x	x/o
BaO·6Fe$_2$O$_3$	o	o	x/o	x/o	o/x
C$_2$H$_5$OH-treated BaO·6Fe$_2$O$_3$	x	o	o/x	o/x	x/o
γ-Fe$_2$O$_3$	o	o	x/o	x/o	o/x
C$_2$H$_5$OH-treated γ-Fe$_2$O$_3$	x	o	o/x	o/x	x/o

[a]o, good dispersion; x, bad dispersion; Δ, little good dispersion; ⊖, collected interface.
Source: Ref. 35.

$$\equiv Si-OH + SiF_4 \longrightarrow \equiv Si-O-Si \overset{F}{\underset{F}{-}} F + HF$$

$$\equiv Si-OH + SO_2C\ell \longrightarrow \equiv Si-C\ell + SO_2 + HC\ell$$

$$\equiv Si-OH + CH_2N_2 \longrightarrow \equiv Si-O-CH_3 + N_2$$

$$\equiv Si-OH + CH_3-CH-CH_2 \longrightarrow \equiv Si-O-CH \overset{CH_3}{\underset{CH_2OH}{<}}$$
$$\underset{O}{\diagdown \diagup}$$

$$\equiv Si-OK + C_2H_5I \longrightarrow \equiv Si-O-C_2H_5 + KI$$

$$\equiv Si-OH + CH_3Li \longrightarrow \equiv Si-O-Li + CH_4$$

Of these, the esterification reaction is used most frequently. The esterification will hardly proceed beyond a certain point, and many studies are being conducted to provide improved methods. Table 5.10 lists observations for silicic acid, alumina, zinc oxide, and a magnetic oxide which were immersed in ethyl alcohol, 1-butyl alcohol or carbon tetrachloride and then heated and pressed in an autoclave, which indicate that the surface of each inorganic filler (powder) became lipophilic by this treatment [35].

5.4 POLYMERIZATION ONTO INORGANIC FILLER SURFACE

5.4.1 Polymerization by Initiator at Filler Surface

If polymerization is induced after allowing an appropriate initiator to be absorbed over the surface of the filler, the end groups of the resulting polymer molecules are connected to the inorganic filler surface. Some reactions have been reported that use such initiators to provide covalent bonding for connecting a graft polymer to an inorganic filler surface [36]. Powder with high fluidity can be produced by allowing methyl methacrylate monomer to react under vacuum with α,α'-azobis(α,γ-dimethylvaleronitrile), an initiator, adsorbed on silicic acid or titanium dioxide.

When azo compounds are adsorbed to clay materials such as kaolin and bentonite, the initiator molecules separate into two groups, one

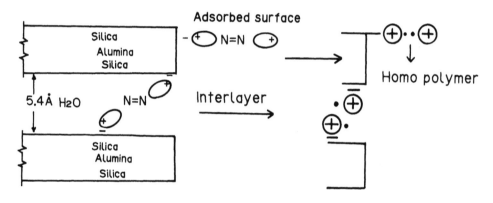

Figure 5.4   Schematic representation of polymerization of monomers initiated by free radicals attached to bentonite. (Ref. 36, 37.)

existing between the layers in the form of a dication and the other on the surface in the form of a monocation, in a ratio of about 80:20 in the case of bentonite. As shown in Figure 5.4, polymerization is initiated mainly between the layers if radicals formed by thermal decomposition coexist with such monomers as methyl methacrylate, styrene, vinyl acetate, chloroprene, and acrylonitrile [37]. For example, polymerization of styrene occurs at a glass surface treated with Lewis acids, including the tetrachloride and titanium tetrachloride. The following complex is likely to be formed in this reaction [38]:

$$-Si-OH + TiCl_4 \longrightarrow -Si-OH \cdot TiCl_4$$

$$\equiv\!Si\!-\!OH\cdot TiCl_4 + CH_2 = CH\!-\!C_6H_5 \longrightarrow \equiv\!Si\!-\!CH_2 - \overset{+}{C}H\!-\!C_6H_5 + TiCl_4OH$$

$$\equiv\!Si\!-\!CH_2 - \overset{+}{C}H\!-\!C_6H_5 + CH_2 = CH\!-\!C_6H_5 \longrightarrow \equiv\!Si\!-\![CH_2 - CH\!-\!C_6H_5\!-\!]_n^{\oplus}$$

Vinyl polymers are produced on a silicic acid surface through similar reactions. The mechanism is as follows:

$$-Si\overset{O}{\underset{}{\diagdown}}Si- \ \xrightarrow{C_6H_5-Li} \ -\underset{OLi}{Si}-O-\underset{C_6H_5}{Si}- \ \xrightarrow{H_2O} \ -\underset{OH}{Si}-O-\underset{C_6H_5}{Si}- \qquad (1)$$

$$-\underset{OH}{Si}- \ \xrightarrow{SOCl_2} \ -\underset{Cl}{Si}- \ \xrightarrow{C_6H_5-Li} \ -\underset{C_6H_5}{Si}- \qquad (2)$$

$$-\underset{O}{Si}- \ \longrightarrow \ -\underset{C_6H_4NO_2}{Si}- \ \longrightarrow \ -\underset{C_6H_4NH_2}{Si}- \ \longrightarrow \ -\underset{C_6H_4N_2^{\oplus}Cl^{\ominus}}{Si}- \qquad (3)$$

$$-\underset{C_6H_4N_2^{\oplus}Cl^{\ominus}}{Si}- \ \xrightarrow{RSNa} \ -\underset{C_6H_4N_2^{\oplus}SR^{\ominus}}{Si}- \ \xrightarrow{\Delta} \ -\underset{C_6H_4\cdot}{Si}- + N_2 \ + SR\cdot \qquad (4)$$

$$-\underset{C_6H_5}{Si}- \ \xrightarrow{nM} \ -\underset{C_6H_5[M]_{\dot{n}}}{Si}- \qquad (5)$$

A phenyl group connected to the silicic acid surface by covalent bonding is formed by reactions (1) and (2). Polymerization is initiated as a result of thermal decomposition of diazonium salts or such derivatives as thioether, which can give radicals. Grafting of styrene, acrylic acid, or vinlypyridine can be performed by these reactions. Twenty to forty percent of the polymer molecules will be grafted to the surface of silicic acid. There are many excellent studies on graft reactions at carbon black surface, including those by Ohkita et al. [39–43]. Some typical graft reactions are shown below.

$$\text{CB} - \text{OH} + \quad \text{CH}_2 = \text{CH}\ (\text{Ph}) \longrightarrow \text{CB} - \text{O} \cdot + \text{CH}_3 - \overset{\cdot}{\text{CH}}\ (\text{Ph}) \tag{6}$$

$$\text{CB} - \text{O} \cdot + \overset{\cdot}{\text{CH}} - \text{CH}_2\ (\text{Ph}) \longrightarrow \text{CB} - \text{O} - \text{CH} - \text{CH}_2 \sim\sim\ (\text{Ph}) \tag{7}$$

$$\text{CB} - \text{COOH} + \text{CH}_2 = \underset{X}{\text{CH}} \longrightarrow \text{CB} - \text{CO}\bar{\text{O}} + \text{CH}_3 - \underset{X}{\overset{+}{\text{CH}}} \tag{8}$$

$$\text{CB} - \text{CO}\bar{\text{O}} + \text{CH}_3 - \underset{X}{\text{CH}} - \left[\text{CH}_2 - \underset{X}{\text{CH}}\right]_{n-2} \text{CH}_2 - \underset{X}{\overset{+}{\text{CH}}} \tag{9}$$

$$\longrightarrow \text{CB} \longrightarrow \underset{O}{\overset{\|}{\text{C}}} - \text{O} - \left[\text{CH} - \underset{X}{\text{CH}_2}\right]_{n-1} \underset{X}{\text{CH}} - \text{CH}_3$$

5.4.2 Copolymerization with Monomer Adsorbed on Inorganic Filler Surface

An acrylic acid/styrene copolymer is formed with strong bonding to the surface of titanium dioxide if acrylic acid or methacrylic acid adsorbed on titanium dioxide is dispersed in styrene monomers and polymerization is initiated by a benzoyl peroxide initiator, as follows [36].

$$\text{C} - \text{CH} = \text{CH}_2 \xrightarrow{\text{St}} \text{C} - \text{CH} - \text{CH}_2 - [\ \text{St}\ ]\overline{\overline{n}}$$

If sodium montmorillonite is used, the compounds can be copolymer-
ized naturally with butadiene, *cis*-2-buten, *trans*-2-butene, or pro-
pionamide without using initiators [44a]. Most of the resulting polymer
molecules are connected to the surface and cannot be extracted [44].
Polymerization of acrylic acid, methyl acrylate, acrylonitrile, 4-vinyl-
pyridine, and styrene between the layers of montmorillonite can begin
without the existence of an initiator. In this reaction, hydrogen
bonding is produced between hydrogen of the monomer and oxygen
on the crystal surface or between oxygen of the former and a hydroxyl group on the latter, leading to a regular arrangement of the
resulting molecules. It is reported that for montmorillonite, monomers
penetrate the gaps between the lattice layers, forming polymer there,
in three different ways, as shown in Figure 5.5 [45].

When hydroxyl groups on an inorganic filler surface are allowed
to react with vinyl isocyanate, vinyl-substituted titanium dioxide or
silicic acid will be formed, which can then undergo a copolymerization reaction with vinyl monomers such as styrene. It is possible to
use such reactions to connect polymer molecules directly to an inorganic filler. For example, polyethylene oxide/isocyanate compounds
are easily added to silica through the following reactions [46]:

$$t\text{-Bu(O-CH}_2\cdot\text{CH}_2)_x\text{OH} + \text{OCNRNCO} \longrightarrow \qquad (10)$$

$$t\text{-Bu(O}\cdot\text{CH}_2\cdot\text{CH}_2)_x\text{OCONHRNCO}$$

$$t\text{-Bu(O}\cdot\text{CH}_2\cdot\text{CH}_2)_x\text{OCONHRNCO} + \text{SiOH} \longrightarrow \qquad (11)$$

$$t\text{-Bu(O}\cdot\text{CH}_2\cdot\text{CH}_2)_x\text{OCONHR-NHCOO-Si}$$

Polymerization can also be initiated at an inorganic filler surface
by irradiation of high-energy light such as γ-rays. Such reactions
have been studied for vinyl chloride, methyl methacrylate, and acrylonitrile on the surface of carbon black or magnesium oxide, and styrene and calcium carbonate on magnesium oxide, as well as styrene
and methyl methacrylate on zinc oxide [36,47]. The mechanism of
the γ-radiation-induced polymerization of methyl methacrylate on a
glass powder surface is interpreted as follows:

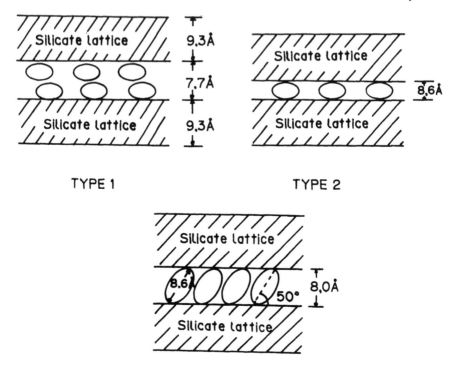

TYPE 1   TYPE 2

TYPE 3

Figure 5.5   Schematic representation of monomer-montmorillonite complexes.   (Ref. 45.)

$$-\overset{|}{\underset{|}{Al}}-OH \xrightarrow{\gamma\text{-ray}} -\overset{|}{\underset{|}{Al}}\cdot \; + \; OH\cdot \tag{12}$$

$$-\overset{|}{\underset{|}{Al}}\cdot \; + \; CH_2\!\!=\!\!\overset{\overset{\displaystyle CH_3}{|}}{\underset{\underset{\displaystyle OCH_3}{|}}{\underset{\displaystyle CO}{|}}{C}} \longrightarrow -\overset{|}{\underset{|}{Al}}\cdot-CH_2-\overset{\overset{\displaystyle CH_3}{|}}{\underset{\underset{\displaystyle OCH_3}{|}}{\underset{\displaystyle CO}{|}}{C}}\cdot \tag{13}$$

$$
\underset{\begin{array}{c}|\\ \end{array}}{-\text{Al}-\text{CH}_2-}\overset{\overset{\displaystyle \text{CH}_3}{|}}{\underset{\underset{\displaystyle \text{OCH}_3}{|}}{\underset{\underset{\displaystyle \text{CO}}{|}}{\text{C}\cdot}}} \quad +n\text{CH}_2=\overset{\overset{\displaystyle \text{CH}_3}{|}}{\underset{\underset{\displaystyle \text{OCH}_3}{|}}{\underset{\underset{\displaystyle \text{CO}}{|}}{\text{C}}}} \quad \longrightarrow \quad -\text{Al}-\text{CH}_2-\overset{\overset{\displaystyle \text{CH}_3}{|}}{\underset{\underset{\displaystyle \text{OCH}_3}{|}}{\underset{\underset{\displaystyle \text{CO}}{|}}{\text{C}}}} \quad -[-\text{CH}_2-\overset{\overset{\displaystyle \text{CH}_3}{|}}{\underset{\underset{\displaystyle \text{O}\cdot \text{CH}_3}{|}}{\underset{\underset{\displaystyle \text{CO}}{|}}{\text{C}}}} -]
$$

(14)

5.5 SURFACE MODIFICATION OF INORGANIC FIBERS

In this section we deal with surface modification and adhesion of inorganic fibers, including glass fiber and carbon fiber, which are widely used for fiber-reinforced composite materials.

5.5.1 Surface Modification by Grafting and Adhesion Properties of Glass Fibers

Silanol and siloxane groups are likely to be the major functional groups on a glass fiber surface since glass consists mainly of SiO_2. It is widely known that when heated, silanol groups decompose into siloxane groups, releasing water in the process, whereas siloxane groups formed at moderate temperatures will rehydrate to form silanol groups in the presence of water. Figure 5.6 shows the results of a study by Davydov et al. [48], who investigated the effects of heat treatment on silanol concentration at a silica surface. Silanol groups existing at a glass surface are connected to water through hydrogen bonding. It has been reported that water molecules adsorbed to silanol groups in a ratio of 1:1 will remain strongly adsorbed even after thorough degassing [49]. Furthermore, the conditions of glass surfaces differ depending on the environment, particularly moisture. There generally exist one or more layers of free water, each having a thickness equal to the molecular diameter, which is weakly adsorbed on the surface. The effects of such adsorbed water on the surface should be taken into account in considering the reactions of silanol and siloxane groups at a glass fiber surface. The reactivity of these functional groups is not discussed here because it was outlined earlier in the chapter and because Boehm's detailed report [50] is available.

The adhesion between the glass fiber and matrix is the most important factor in relation to the properties of a glass fiber/polymer composite. Incomplete adhesion will cause deterioration of the mechanical, electrical, and moisture-resistant properties of the composite.

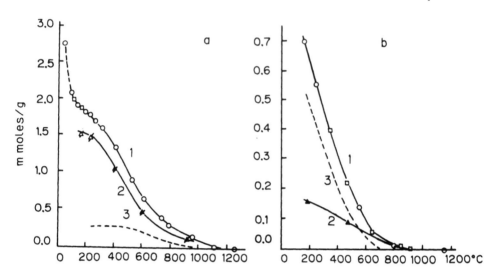

Figure 5.6   Curves 1, total water loss due to thermal treatment of
hydroxylate silica gels VI(a) and VII(b). ○ and □, measured values.
Curves 2, content of surface structural water for silica gels VI

(⊬—Si—OH + D_2O and ⊬—Si—OD + H_2O, exchange at 150°; (△—Si—

—OH + D_2O, exchange at 20°C) and VII (▲—Si—OH + D_2O), de-

pending on the temperature of treatment. Curves 3, content of inner structural water of silica gels VI and VII (obtained by difference of ordinates corresponding to curves 1 and 2). (Ref. 48.)

Laird et al. [51] have reported that in fiber-reinforced polymer, the
diffusion rate of water along the interface between the glass fiber
and polymer is about 450 times greater than that through the resin
layers, indicating the complete adhesion of glass fiber and polymer
is highly important.
        In general, the critical surface tension, γ_c, at a chemically clean
glass surface is greater than 80 dyn/cm, while it decreases to 20 to
40 dyn/cm when the surface is stained with organic material. Moreover, it can decrease to 28 and 46 dyn/cm, respectively, in an atmosphere at relative humidity 95% and 1%, even if there is no organic contamination (Table 5.11) [52]. Wetting of glass fiber by liquid
polymer tends to be incomplete and voids are often formed since the
liquid polymers used for composites usually have a surface tension

Table 5.11   Spreading Behavior on Glass of Pure Liquids Having Negative and Positive Spreading Coefficients on Bulk Water[a]

Liquid sessile drop	Surface tension, γ_{LV} (dyn/cm)	Contact angle θ (deg), or spreading behavior of drop on glass equilibrated at:		
		95% RH	65–53% RH	1% RH
Methylene iodine	50.8	36	31	13
Tetrabromoethane	47.5	36	9	9
1-Methylnaphthalene	38.7	7	–	<5
Isopropyl biphenyl	34.8	16	–	<5
Dicyclohexyyl	32.8	21	21	Spread
p-Octadecyl toluene	31.5	17	–	<5
Isopropyl bicyclohexyl	30.9	13	–	<5
Squalane	28.5	Spread	–	Spread
n-Hexadecane	27.6	Spread	Spread	Spread
Water	72.8	Spread	–	–
Formamide	58.2	Spread	–	–
Thiodiglycol	54.0	Spread	–	Spread
Ethylene glycol	47.7	Spread	–	–
Propylene carbonate	41.1	Spread	–	Spread
Tricresyl phosphate	40.9	6	–	14
Hexachloropropylene	38.1	Spread	–	Spread
Bis(2-ethylhexyl) phthalate	31.3	23	–	–
Bis(2-ethylhexyl) sebacate	31.1	17	–	–
Pentaerythyritol tetracaproate	30.4	8	–	–

[a]All data at 20°C.
Source: Ref. 52.

of 50 to 90 dyn/cm [53]. The method used most widely to solve this problem is to treat the glass fiber with coupling agents such as those mentioned previously. We will not discuss this further here as many reports are available in the literature.

There are few publications that deal with graft polymerization onto glass fiber surfaces. However, it has been reported that unsaturated bonds can be introduced through the reaction of a silanol group and allyl glycidyl ether, as shown below, where graft polymerization proceeds from ordinary vinyl monomers and an initiator [54].

$$
\begin{array}{c}
\text{OH} \\
|\ \\
-\text{Si}- \ +\ \underset{\underset{\text{O}}{\diagdown\diagup}}{\text{CH}_2-\text{CH}}-\text{CH}_2-\text{O}-\text{CH}_2-\text{CH}=\text{CH}_2\ \longrightarrow
\end{array}
\tag{15}
$$

$$
\begin{array}{c}
\text{O}-\text{CH}_2-\text{CH}-\text{CH}_2-\text{O}-\text{CH}_2-\text{CH}=\text{CH}_2 \\
|\qquad\qquad| \\
-\text{Si}-\qquad\ \text{OH}
\end{array}
$$

$$
\begin{array}{c}
\text{O}-\text{CH}_2-\text{CH}-\text{CH}_2-\text{O}-\text{CH}_2-\text{CH}=\text{CH}_2\qquad\longrightarrow \\
|\qquad\qquad| \\
-\text{Si}-\qquad\ \text{OH}\qquad\qquad\qquad\qquad +\ \text{R}\cdot
\end{array}
\tag{16}
$$

$$
\begin{array}{c}
\text{O}-\text{CH}_2-\text{CH}-\text{CH}_2-\text{O}-\text{CH}-\text{CH}=\text{CH}_2 \\
|\qquad\qquad| \\
-\text{Si}-\qquad\ \text{OH}\qquad\qquad\qquad\qquad +\ \text{RH}
\end{array}
$$

$$
\begin{array}{c}
\text{O}-\text{CH}_2-\text{CH}-\text{CH}_2-\text{O}-\overset{\displaystyle\cdot}{\text{C}}\text{H}-\text{CH}=\text{CH}_2 \\
|\qquad\qquad| \\
-\text{Si}-\qquad\ \text{OH}\qquad\qquad\qquad +\ n\text{CH}_2{=}\text{CH}\ \longrightarrow \\
\qquad\qquad\qquad\qquad\qquad\qquad\qquad\ | \\
\qquad\qquad\qquad\qquad\qquad\qquad\qquad\ \text{X}
\end{array}
\tag{17}
$$

$$
\begin{array}{c}
\qquad\qquad\qquad\qquad\text{CH}=\text{CH}_2 \\
\qquad\qquad\qquad\qquad\ | \\
\text{O}-\text{CH}_2-\text{CH}-\text{CH}_2-\text{O}-\text{CH}-\!\!\left(\!\text{CH}_2-\text{CH}\!\right)_{\!n} \\
|\qquad\qquad|\qquad\qquad\qquad\qquad\quad| \\
-\text{Si}-\qquad\ \text{OH}\qquad\qquad\qquad\qquad\text{X}
\end{array}
$$

Table 5.12 shows the properties of a composite consisting of an epoxy resin and glass cloth treated by the method described above. These data suggest that the interlaminar shear strength (i.e., the adhesion strength between the glass fiber and the matrix resin) increases with increasing graft ratio.

Table 5.12 Mechanical Properties of Grafted Glass Cloth-reinforced Epoxy Resin Composites[a]

Glass cloth treatment[b]		Flexural strength [(N/m^2) × 10^{-7}]	Flexural modulus [(N/m^2) × 10^{-7}]	Interlaminar shear strength [(N/m^2) × 10^{-7}]
None		53.4 (52.0)	2479 (2355)	4.8 (3.8)
MMA-grafted	(a)	57.8 (53.0)	2508 (2408)	5.4 (4.0)
	(b)	58.2 (54.4)	2530 (2440)	5.8 (4.0)
	(c)	58.0 (57.2)	2558 (2515)	6.0 (5.6)
St-grafted	(d)	56.8 (53.1)	2510 (2410)	5.2 (3.9)
	(e)	58.0 (54.8)	2574 (2430)	5.8 (4.1)
	(f)	60.2 (59.4)	2635 (2594)	6.2 (6.0)
PolyMMA-coated	(g)	55.3 (52.4)	2485 (2360)	5.0 (3.8)
PolySt-coated	(h)	55.6 (52.5)	2490 (2370)	5.0 (3.8)

[a]Values in parentheses are for boiling test for 72 h, at 100°C.

[b](a) Grafting 3.2%; (b) grafting 9.5%; (c) MMA-MAA (90:10) grafted; grafting 4.2%; (d) grafting 2.8%; (e) grafting 8.6%; (f) St-GMA (95:5) grafted; grafting 5.6%; (g) coating 7.5 wt % to a hydrothermal-treated glass cloth; (h) coating 8.2 wt % to a hydrothermal-treated glass cloth.

Source: Ref. 54.

If glass fiber treated with thionyl chloride is allowed to react with 2-mercaptoethanol to form mercapto groups on the fiber surface, the resulting surface can undergo radical graft polymerization with styrene or methyl methacrylate, indicating that the reaction can be used to modify glass fiber surfaces [55]. It is also possible to introduce isocyanate groups onto a fiber surface by allowing glass fibers to react with diisocyanate [56,57]. The use of the treated fiber leads to a large increase in adhesion strength at the fiber/matrix resin interface in glass fiber-reinforced rigid urethane foam [58,59]. There are other reports that deal with graft polymerization of styrene, acrylonitrile, or phenylacetylene-acrylonitrile onto glass fiber surfaces with the aid of x-ray or other radiations [60].

In addition to these grafting techniques, glass fiber can be encupsulated with fibrils running parallel to each other if copolymerization of styrene and acrylonitrile is performed in water containing dispersed chopped strands, resulting in a higher degree of mechanical strength and less water absorption than in ordinary blend systems [61].

5.5.2 Surface Modification and Adhesion Properties of Carbon Fibers

Carbon fibers are generally produced from polyacrylonitrile (PAN) fiber, liquid crystal pitch fiber, or rayon fiber. They are grouped into general-purpose type and high-performance type, depending on the starting raw material used, and the latter is subgrouped into high-strength (HS) type and high-modulus (HM) type. As carbonized carbon fiber products have relatively high wettability by matrix resins and show rather high reinforcing effects without the aid of coupling agents, whereas a product with a higher modulus tends to be less easily wetted by matrix resins, showing a lower interlaminar shear strength value.

Table 5.13 lists contact angles for various carbon fiber products with liquids [62]. It can be seen that the properties of a carbon fiber surface have a considerable effect on the strength of the resulting composite. Various types of surface treatment have been used to improve adhesion at the interface between fiber surfaces and matrix resins. The most typical method used is the introduction of an oxygen-containing functional group by oxidation treatment. Both the type of carbon fiber and the process and conditions for oxidation treatment affect this method significantly. Generally, air oxidation at a high temperature acts to create small pores and etch pits, although it is difficult to introduce functional groups by this process. This method can be carried out relatively easily by wet oxidation techniques using (1) sulfuric acid, (2) potassium permanganate/sulfuric acid, or (3) sodium hypochlorite [63,64]. The anodic oxidation technique [65], in which electrolysis is carried out in a solution of

Table 5.13 Wettability of Various Carbon Fibers

Carbon fiber	Treatment	Liquid	Contact angle, θ (deg)
Thermolon S–M (PAN based HM)	None	Epoxy araldite	72
	Oxidated with 60% HNO$_3$ at 70°C for 1 h	⎛ MY750 ⎞ ⎜ MNA ⎟ ⎝ BDMA ⎠	66
Thermolon S–T (PAN based HS)	None		69
	Oxidated with 60% HNO$_3$ at 70°C for 1 h		61
Thornel 125 (Rayon based HM)	None	H$_2$O	36
	–	Epoxy	
		ERLA-0400	12 ± 7
		ERL-2774	32 ± 8
		H$_2$O	38 ± 8
	Oxidated with O$_2$ (2 ml/min) at 500°C for 15 min	Epoxy ERLA-0400	4.3 ± 0.6
		ERL-2774	5.4 ± 0.7

Source: Ref. 62.

sulfuric acid or sodium hydroxide is also helpful. Polar groups such as carboxyl (COOH), carbonyl (CO), and phenolic hydroxyl groups are introduced by these processes. This surface structure model is considered to be similar to that proposed by Boehm et al. [50] for oxides formed on the surface of activated charcoal, shown in Figure 5.7 [65].

The introduction of these polar groups serves to increase the wettability and adhesion strength between the fiber and matrix resins, leading to a great increase in interlaminar shear strength [66,67], as can be seen from the measurements given in Figure 5.8. Furthermore, it has been reported that surface treatment is carried out most effectively by combining a nitric acid oxidation process with additional heat treatment to provide a clean, rough fiber surface [68].

Many techniques have been studied to improve the properties of composites by making polymers form chemical bonds to the surface of carbon fiber. It has been demonstrated that for various vinyl monomers, graft copolymerization can be induced by an initiator if an unsaturated group [69], mercapto group [70], or perester group

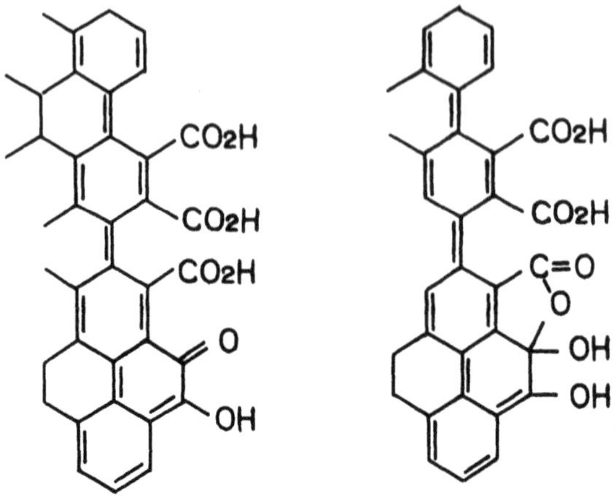

1 open form   2 lactone form

Figure 5.7   Boehm models of the oxidated carbon fiber surface.
(Ref. 65.)

[71] has been introduced on the carboxyl group formed by the nitric acid oxidation treatment. The wettability, tensile stress, and breaking extension of carbon fiber are increased with increasing graft ratio, which leads to a large increase in flexural strength and interlaminar shear strength in epoxy resin composites [72].

It has also been reported that if carbon fiber with an isocyanate group is combined with rigid polyurethane, the resulting composite will have improved flexural properties and increased interlaminar shear strength and, in addition, these properties will deteriorate only slightly during hydrothermal treatment, as can be seen from the measurements given in Table 5.14 [73]. Other studies have shown that a carboxylated polymer or soft polymer can be grafted to ozone-treated carbon fiber using γ-radiation [74] and that the interlaminar shear strength and impact resistance of composites are increased by isoprene-styrene graft copolymerization [75].

Graft polymerizations based on electric conductivity, a unique function of carbon fibers, were developed by Subramanian et al. [76], who carried out electrolytic polymerization for various vinyl monomers and multifunctional monomers in electrolite solutions to form graft polymers, listed in Table 5.15. It has been suggested that in these reactions, radicals on carbon fiber surfaces formed from acid groups with the aid of electricity will polymerize or react with growing radical chains to produce graft polymers. Characterized by extremely

high rates of reactions, which tend to be completed in a few seconds, this method has promise for use in a continuous treatment
process during carbon fiber manufacture. If carbon fiber produced by this method is used with an epoxy resin, the resulting
composite will have either increased interlaminar shear strength or
increased impact resistance, depending on the type of monomer used
for grafting, as shown in Figures 5.9 and 5.10. Although these
effects are generally antagonistic to each other, the aziridinyl monomer shown below (PFZ-3000) has been found to act exceptionally
well to improve both of these properties.

$$R - \left(N \begin{array}{c} {\diagup} CN_2 \\ | \\ {\diagdown} CN_2 \end{array} \right)_3 \qquad R = \text{long-chain alkyl group}$$

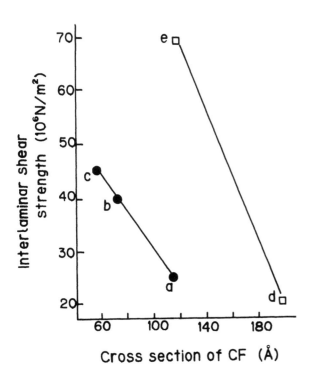

Figure 5.8   Relation between interlaminar shear strength of composite and a cross section of carbon fiber.   a, Thornel 125; b,
Thornel 125 (air oxidation); c, Thornel 125 (oxidation by HNO_3);
d, Morganite I; e, Morganite I (oxidation by HNO_3).   (Ref. 67.)

Table 5.14   Bending Properties of Carbon Fiber-Reinforced Rigid Polyurethane Foam[a]

Carbon fiber treatment	Carbon fiber content (vol %)	Flexural strength (kg/mm^2)	Flexural modulus (kg/mm^2)	Interfacial shear strength (kg/mm^2)
Epoxy resin	5.8	11.4	387	0.85
treatment		(10.2)	(375)	(0.45)
	3.3	9.6	355	0.84
		(9.2)	(346)	(0.44)
	1.6	8.4	307	0.74
		(7.0)	(285)	(0.38)
Oxidation	5.8	12.5	390	0.89
followed by		(12.0)	(384)	(0.88)
diisocyanate	3.3	10.4	362	0.88
treatment		(9.8)	(352)	(0.86)
	1.6	8.8	315	0.80
		(8.5)	(308)	(0.75)

[a]Values in parentheses are for samples after boiling-water treatment for 30 h.
Source: Ref. 73.

Table 5.15   Polymer Grafting to Graphite Fiber by Electropolymerization

System	Electrode	Grafting observed
DAA, $H_2SO_4-H_2O$	Cathode	Yes
Methyl methacrylate, $NaNO_3-DMF$	Cathode	Yes
Styrene, $NaNO_3-DMF$	Cathode	Yes
Methyl methacrylate, $LiOAc-CH_3OH$	Cathode	Yes
Acrylic acid, $H_2SO_4-H_2O$	Cathode	Unclear
Acrylic acid, $H_2SO_4-H_2O$	Anode	No
Methyl methacrylate/acrylic acid, LiOAc-EtOH	Anode	Yes
Methyl methacrylate, $NaNO_3-DMF$	Anode	Yes
ε-Caprolactam, $NaNO_3-DMF$	Cathode	No
Acrylonitrile, $(CH_3)_4NCl$	Cathode	No
Vinyl acetate, $LiOAc-CH_3OH$	Anode	No

Source: Ref. 76.

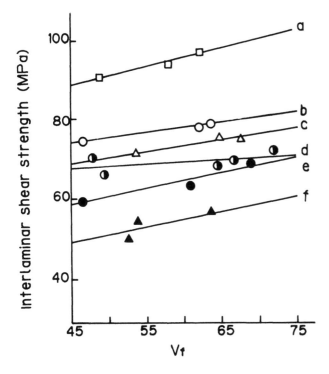

Figure 5.9 Shear strength of composites prepared from graphite fibers. a, Hercules AS and Hercules AU coated by electropolymerization of various monomers; b, acrylic acid; c, DAA; d, untreated Hercules AU; e, MMA; f, styrene-acrylonitrile. 2.5 s polymerization time. (Ref. 76.)

This electrolytic polymerization method is also effective for polyimide intermediates to form heat-resistant carbon fiber composites. In addition, studies have been conducted to ionize acids or polymers with anhydrous acid radicals in water and electrodeposit them on carbon fiber used as the anode. Graft polymers can be produced from polymer radicals formed by anodic oxidation in the Kolbe reaction. The mechanical properties of composites are improved by combining these carbon fibers with epoxy resins, as shown in Table 5.16 [77].

5.5.3 Surface Modification and Adhesion Properties of Other Inorganic Fibers

Adhesion properties at the interface have been studied for various composites consisting of various matrix resins and fibrous inorganic

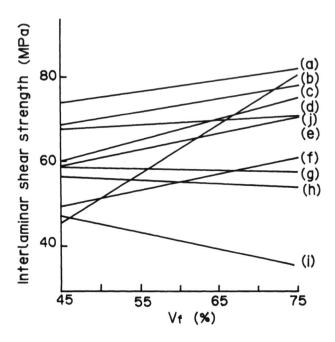

Figure 5.10   Shear strength of composites prepared from Hercules AU graphite fibers coated by electropolymerization. a, Acrylic acid; b, PFAZ 300; c, DAA; d, styrene; e, MMA; f, styrene-acrylonitrile; g, ε-caprolactam; h, Epon 828/phthalic anhydride; i, VTBN; j, un-treated Hercules AU. Data points omitted for clarity. 2.5 s polymerization time. (Ref. 76.)

materials, including asbestos [78], xonotlite (needlelike crystal of calcium cilicate) [79], wollastonite (needlelike white natural mineral) [80], fibrous gypsum [81], and attapulgite (fibrous clay) [82], treated with silane coupling agents. Polypropylene/thermoplastic elastomer blend composite filled with calcium carbonate modified with mono-alkyl phosphate esters showed major improvements in modulus and impact properties [83]. Graft polymers were reportedly formed on asbestos by radiation-induced polymerization of styrene and methyl methacrylate. Furthermore, radical polymerization of vinyl monomers in water-containing dispersed asbestos leads to encapsulation of the fiber surface with the resulting polymer. Composites composed of such fiber materials have highly increased toughness and impact resistance as well as improved moldability [84].

To improve the heat resistance, strength, and elastic modulus of composites, ceramic fibers or whiskers of silicon carbide, alumina, silicon nitride, boron nitride, zirconia, and potassium titanate have

Table 5.16 Effect of Electrodeposited Polymer on Graphite Fiber Composite Properties (V$_f$: 50%)[a]

Electrodeposited polymer[b]	Interlaminar shear strength (MPa)	Impact strength (kJ/m^2)	Flexural strength (MPa × 10^{-1})
SMA 1000	68	57	110
SMA 2000	59	72	110
SMA 3000	62	56	100
PA 6	61	42[c]	100
PA 18	52	44[c]	91
AN 119	48	86	90
AN 139	59	130	95
AN 169	54	140	86
None (CG-3)[d]	34	63	78
None (CG-3)[e]	52	43	96

[a]Fortafil fiber, (170 GPa modulas); epoxy matrix (diglycidyl ether of bisphenol-A cured by methaphenylenediamine).

[b]Copolymers of maleic anhydride with styrene (SMA), hexene (PA 6), octadecene (PA 18), and methyl vinyl ether (AN); described fully in the text.

[c]Fibers electrocoated in modified apparatus.

[d]Untreated, Fortafil CG-3 fiber.

[e]Surface-treated Fortafil CG-3 fiber.
Source: Ref. 77.

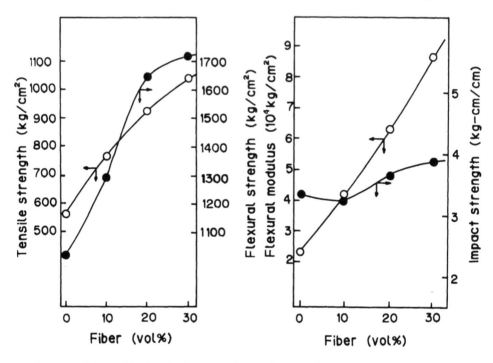

Figure 5.11   Mechanical properties of potassium titanate fiber-
reinforced polyacetal resin.   (Ref. 85.)

recently been developed and studies are being conducted in their
use as reinforcing materials.   Metals or ceramics are employed as
matrices in most of publications that address composites consisting
of these substances.   There are extremely few studies on polymer
matrices, including only two or three that deal with the production
of composites using polymers with surfaces treated with silane or
titanate coupling agents, as shown in Figure 5.11 [85–87].
     Recently, Inubushi et al reported that the mica flakes surface,
treated with a combination of epoxy prepolymer, reinforced thermoplastics, and various aminimide curing agents, as shown in Figure
5.12, significantly improved the filler-matrix interface [88,89].

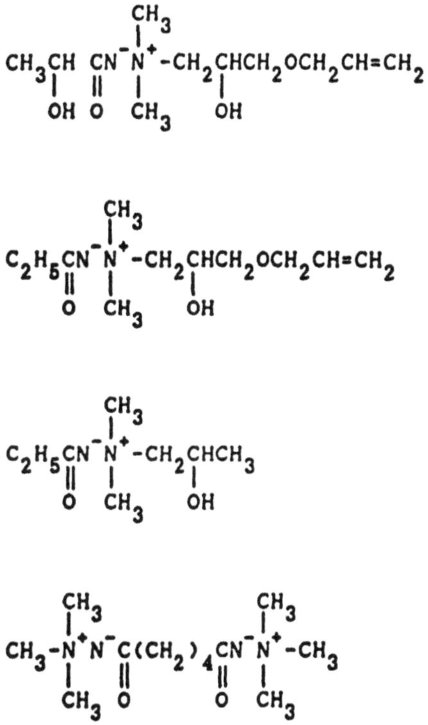

$$CH_3CH\underset{\underset{OH}{|}}{CN}\underset{\underset{O}{\|}}{-}\overset{\overset{CH_3}{|}}{N}\overset{+}{}\underset{\underset{CH_3}{|}}{-}CH_2\underset{\underset{OH}{|}}{CH}CH_2OCH_2CH=CH_2$$

$$C_2H_5\underset{\underset{O}{\|}}{CN}\overset{\overset{CH_3}{|}}{N}\overset{+}{}\underset{\underset{CH_3}{|}}{-}CH_2\underset{\underset{OH}{|}}{CH}CH_2OCH_2CH=CH_2$$

$$C_2H_5\underset{\underset{O}{\|}}{CN}\overset{\overset{CH_3}{|}}{N}\overset{+}{}\underset{\underset{CH_3}{|}}{-}CH_2\underset{\underset{OH}{|}}{CH}CH_3$$

$$CH_3-\overset{\overset{CH_3}{|}}{N}\overset{+}{}\underset{\underset{CH_3}{|}}{N}\overset{-}{}\underset{\underset{O}{\|}}{C}(CH_2)_4\underset{\underset{O}{\|}}{CN}\overset{-}{}\overset{\overset{CH_3}{|}}{N}\overset{+}{}\underset{\underset{CH_3}{|}}{-}CH_3$$

Figure 5.12   Structures of aminimide compounds.   (Ref. 88, 89.)

REFERENCES

1. Catalogs of Shinetsu Silicone Ltd. and Torey Silicone Ltd.

2. T. Nakao, The Use of Titanate Coupling Agent, *Finechem. Jpn.*, 4, 3, (1984).

2a. K. Kubo, M. Koishi, and T. Tsunoda, (eds.) *Composite Materials and Interface*, Sogo Gijutsu Shuppan, Tokyo (1986).

3. H. Ishida and J. L. Koenig, Fourier Transform Infrared Spectroscopic Study of the Silane Coupling Agent/Porous Silica Interface, *J. Colloid Interface Sci.*, 64, 555 (1978).

4. S. Wu, *Polymer Interface and Adhesion*, Marcel Dekker, Inc., New York, 1982.

5. J. G. Vail, *Soluble Silicate*, Reinhold Publishing Corporation, New York, p. 171.

6. B. M. Vanderbilt, Effectiveness of Coupling Agents in Glass-Reinforced Plastics, *Mod. Plast.*, 37, 125 (1959).

7.   H. A. Clark and E. P. Plueddemann, Bonding of Silane Coupling Agents in Glass-Reinforced Plastics, *Mod. Plast.*, *40*, 133 (1963).

8.   J. L. Koenig and P. T. K. Shih, Raman Studies of the Glass Fiber-Silane-Resin Interface, *J. Colloid Interface Sci.*, *36*, 247 (1971).

9.   R. L. Kaas and J. L. Kardos, The Interaction of Alkoxy Silane Coupling Agents with Silica Surface, *Polym. Eng. Sci.*, *11*, 11 (1971).

10. P. T. K. Shih and J. L. Koenig, Raman Studies of Silane Coupling Agents, *Mater. Sci. Eng.*, *20*, 145 (1975).

11. F. J. Boerio and S. Y. Cheng, Adsorption of γ-Methacryloxy-propyltrimethoxysilane onto Copper, *J. Colloid Interface Sci.*, *68*, 252 (1979).

12a. P. T. K. Shih and J. L. Koenig, Raman Studies of the Hydrolysis of Silane Coupling Agents, *Mater. Sci. Eng.*, *20*, 137 (1975).

12b. H. Ishida, S. Naviroj, S. K. Tripathy, J. J. Fitzgerald, and J. L. Koenig, The Structure of an Aminosilane Coupling Agent in Aqueous Solutions and Partially Cured Solids, *36th Annu. Tech. Conf. SPI*, 2-C (1981).

13. F. Ide, Application of Graftcopolymerization and Block Copolymer, *Plast. Jpn.*, *32*, 30 (1981).

14a. C. Chiang and J. L. Koenig, Chemical Reactions Occurring at the Interface of Epoxy Matrix and Aminosilane Coupling Agents in Fiber-Reinforced Composite, *35th Annu. Tech. Conf. SPI*, 23-D (1980).

14b. E. P. Plueddemann, Adhesion Through Silane-Coupling Agents, *25th Annu. Tech. Conf. SPI*, 13-D (1970).

15. W. A. Zisman, Improving the Performance of Reinforced Plastics, *Ind. Eng. Chem.*, *57*, 26 (1965).

16. W. A. Zisman, Surface Chemistry of Plastics Reinforced by Strong Fibers, *Ind. Eng. Chem. Prod. Res. Dev.*, *8*, 98 (1969).

17. P. W. Erickson, Historical Background of the Interface: Studies and Theories, *25th Annu. Tech. Conf. SPI*, 13−A (1970).

18. E. P. Plueddemann, Cationic Silane Coupling Agents for Thermoplastics, *27th Annu. Tech. Conf. SPI*, 11−B (1972).

19. E. P. Plueddemann and G. L. Stark, Adhesion to Silane-Treated Surfaces Through Interpenetrating Polymer Networks, *35th Annu. Tech. Conf. SPI*, *20-B* (1980).

20. Y. L. Fan and R. G. Shaw, Adesion Promotion with Silylper-oxides, *Mod. Plast.*, *5*, 104 (1970).

21. Y. L. Fan and R. G. Shaw, Silylperoxides—A Novel Family of Adhesion Promoters, *25th Annu. Tech. Conf. SPI*, *16-A* (1970).

22. S. Inubushi, T. Ikeda, and S. Tazuka, Aminimide-cured Epoxy Resins as Surface Modifiers for Mica Flakes in Particle-reinforced Thermoplastics, *J. Mater. Sci.*, *23*, 535, (1988).

23. S. J. Monte and G. Sugerman, A New Generation of Age and Water Resistant Reinforced Plastics, *Polym. Plast. Technol. Eng.*, *13*, 115 (1979).

24. C. D. Han, C. Sandford, and H. J. Yoo, Effect of Titanate Coupling Agents on the Rheological and Mechanical Properties of Filled Polyolefins, *Polym. Eng. Sci.*, *18*, 849 (1978).

25. E. Tsuchiya, and M. Takehara, Studies on Surface Modifications of Inorganic Fillers Using Titanate Coupling Agents I, *J. Soc. Colour Matel.*, Jpn., *57*, 363 (1984).

26a. D. Rivin, Use of Lithium Aluminum Hydride in the Study of Surface Chemistry of Carbon Black, *Rubber Chem. Technol.*, *36*, 729 (1963).

26b. G. R. Cotten, B. B. Boonstra, D. Rivin, and F. R. Williams, Effect of Chemical Modification of Carbon Black on Its Behavior in Rubber, *Kautsch. Gummi*, *22*, 477 (1969).

27. Y. Hirata, Structure of Filler and Compound Properties, *J. Soc. Rubber Ind. Jpn.*, *45*, 625 (1972).

28. H. P. Boehm, Funktionelle Gruppen an Festkörper-Oberflächen, *Angew. Chem.*, *78*, 617 (1966).

29. D. J. C. Yates, Infrared Studies of the Surface Hydroxyl Groups on Titanium Dioxide, and of the Chemisorption of Carbon Monoxide and Carbon Dioxide, *J. Phys. Chem.*, *65*, 746 (1961).

30. I. T. Smith, Infrared Spectra of Polar Molecules Adsorbed on Titanium Dioxide Pigments, *Nature*, *201*, 67 (1964).

31. J. M. Mays and G. W. Brady, Nuclear Magnetic Resonance Absorption by H_2O on TiO_2, *J. Chem. Phys.*, *25*, 583, (1956).

32. M. L. Hair, *Infrared Spectroscopy in Surface Chemistry*, Marcel Dekker, Inc., New York 1967.

33. T. Morimoto and M. Nagao, Rate of Surface Hydration on Alumina, *Kolloid. Z. Z. Polym.*, *224*, 62 (1968).

34. K. Taniguchi, M. Nakajima, S. Yosida, and K. Tarama, Ex-Change on the Surface of Silicagel by Metal Ion, *J. Chem. Soc. Jpn.*, *91*, 525 (1970).

35. H. Utsugi, A. Watanabe, and S. Nishimura, Preparation of Lipophilic Metaloxide and Its Properties, *J. Chem. Soc. Jpn.*, *91*, 431 (1970).

36. R. Kroker, M. Schneider, and K. Hamann, Polymer Reactions on Powder Surfaces, *Prog. Org. Coat.*, *1*, 23 (1972).

37. H. G. G. Dekking, Propagation of Vinyl Polymers on Clay Surfaces, *J. Appl. Polym. Sci.*, *11*, 23 (1967).

38. T. E. Lipatova and I. S. Skorynina, Grafting Some Polymers on a Glass Surface Treated with Titanium Tetrachloride, *J. Polym. Sci.*, *C-16*, 2341 (1967).

39. K. Ohkita, N. Tsubokawa, and E. Saitoh, The Competitive Reactions of Initiator Fragments and Growing Polymer Chains Against the Surface of Carbon Black, *Carbon*, *16*, 41 (1978).

40. N. Tsubokawa, Cationic Polymerization of α-Methylstyrene Initiated by Channel Black Surface, *J. Polym. Sci. Polym. Lett. Ed.*, *18*, 461 (1980).

41. N. Tsubokawa, N. Takeda, and A. Kanamaru, The Cationic Polymerization of N-Vinyl-2-pyrrolidone Initiated by Carbon Black Surface, *J. Polym. Sci. Polym. Lett. Ed.*, *18*, 625 (1980).

42. K. Ohkita, N. Nakayama, and M. Shimomura, The Polymerization of Styrene Catalyzed by n-Butyllithium in the Presence of Carbon Black, *Carbon*, *18*, 277 (1980).

43. M. Shimomura, Y. Sanada, and K. Ohkita, Reaction of Growing Polymer Cations on Carbon Black Surface, *Carbon*, *19*, 326 (1981).

44a. H. Z. Friedlander and C. R. Frink, Organized Polymerization III, Monomers Intercalated in Montmorillonite, *J. Polym. Sci. Polym. Lett. B2*, 475 (1964).

44b. A. Blumstein, Polymerization of Adsorbed Monolayers: I. Preparation of the Clay-Polymer Complex, *J. Polym. Sci.*, *A-3*, 2653 (1965).

44c. A. Blumstein, Polymerization of Adsorbed Monolayers: II. Thermal Degradation of the Inserted Polymer, *J. Polym. Sci.*, *A-3*, 2665 (1965).

45. D. H. Solomon and B. C. Loft, Reactions Catalyzed by Minerals: III. The Mechanism of Spontaneous Interlamellar Polymerizations in Aluminosilicates, *J. Appl. Polym. Sci.*, *12*, 1253 (1968).

46. K. Bridger and B. Vincent, The Terminal Grafting of Polyethyleneoxide Chains to Silica Surfaces, *Eur. Polym. J.*, *16*, 1017 (1980).

47. T. Ihara, S. Ito, and T. Kuwahara, Radiation Graft Polymerization of Styrene onto γ-Alumina and Anodic Oxide Film on Aluminum, *Met. Sur. Technol. Jpn*, *29*, 529 (1978).

48. V. Y. Davydov, A. V. Kiselev, and L. T. Zhuravlev, Study of the Surface and Bulk Hydroxyl Groups of Silica by Infrared Spectra and D_2O-Exchange, *Trans. Faraday Soc.*, *60*, 2254 (1964).

49. W. Stöber, Adsorptionseigenschaften and Oberflächenstruktur von Quarzpulvern, *Kolloid. Z.*, *145*, 17 (1956).

50. H. P. Boehm, Chemical Identification on Surface Groups, *Adv. Catal.*, *16*, 179 (1966).

51. J. A. Laird and F. W. Nelson, The Effect of Glass Surface Chemistry on Glass-Epoxy Systems, *SPE Trans*, 120 (1964).

52. E. G. Shafrin and W. A. Zisman, Effect of Adsorbed Water on the Spreading of Organic Liquids on Soda-Lime Glass, *J. Am. Ceram. Soc.*, *50*, 478 (1967).

53. E. J. Kohn, A. G. Sands, and R. C. Clark, Quantitative Measurement of Void Content in Glass-Filament-Wound Composites and Correlation of Interlaminar Shear Strength with Void Content, *Ind. Eng. Chem. Res. Dev.*, *7*, 179 (1968).

54. K. Hashimoto, T. Fujisawa, M. Kobayashi, and R. Yosomiya, Graft Copolymerization of Glass Fiber and Its Application, *J. Appl. Polym. Sci.*, *27*, 4529 (1982).

55. T. Fujisawa, M. Kobayashi, K. Hashimoto, and R. Yosomiya, Graft Copolymerization of Glass Fiber Having Pendent Mercaptan Groups and Its Application, *J. Appl. Polym. Sci.*, *27*, 4849 (1982).

56. R. Yosomiya and K. Morimoto, Glass Fiber Having Isocyanate Groups on the Surface, *Polym. Bull.*, *12*, 41 (1984).

57. R. Yosomiya, K. Morimoto, and T. Suzuki, The Reaction of Glass Fiber with Diisocyanate and Its Application, *J. Appl. Polym. Sci.*, *29*, 671 (1984).

58.   R. Yosomiya and K. Morimoto, Compressive Properties of Glass
      Fiber Reinforced Rigid Polyurethane Foams, *Ind. Eng. Chem.
      Prod. Res. Dev.*, *23*, 605 (1984).

59.   K. Morimoto, T. Suzuki, and R. Yosomiya, Adhesion Between
      Glass Fiber and Matrix of Glass Fiber Reinforced Rigid Poly-
      urethane Foam Under Tension, *Polym. Plast. Technol. Eng.*,
      22, 55 (1984).

60a. A. V. Vlasov, P. Ia, Glaznov, Iu. L. Morozov, I. I. Patalax,
      and L. S. Polak, Synthesis of Materials Combined with Semi-
      conductor by Gas-Phase Radiation Graft Copolymerization, *Dokl.
      Akad. Nauk SSSR*, *158*, 141 (1964).

60b. N. Okada, Ultraviolet Ray and Electron Beam Curing and Ad-
      hesion, *J. Soc. Adhes. Jpn.*, *18*, 307 (1982).

61.   M. Baer, Composites Obtained by Encapsulation and Collimation
      of Glass Fiber Within a Thermoplastic Matrix by Means of Poly-
      merization, *J. Appl. Polym. Sci.*, *19*, 1323 (1975).

62.   S. Otani, Surface Structure and Surface Property of Carbon
      Fiber, *Surf. Jpn.*, *11*, 625 (1973).

63.   J. C. Goan, L. A. Joo, and G. E. Sharpe, Surface Treatment
      for Graphite Fibers, *27th Annu. Tech. Conf. SPI*, *21-E* (1972).

64.   J. B. Donnet, E. Papirer, and H. Dauksch, Surface Modifica-
      tion of Carbon Fibers and Their Adhesion to Epoxy Resins,
      Int. Conf. Carbon Fiber, London, pap. 9 (1974).

65.   N. L. Weinberg and T. B. Reddy, Electrochemical Oxidation
      of the Surface of Graphite Fibers, *J. Appl. Electrochem.*, *3*,
      73 (1973).

66a. S. Otani, M. Tamura, Y. Yoshida, and Y. Tsuji, The Prepara-
      tion and General Properties of Ion-Exchange Carbon Fiber, *Bull.
      Chem. Soc. Jpn*, *45*, 1908 (1972).

66b. B. Rand and R. Robinson, A Preliminary Investigation of PAN
      Based Carbon Fiber Surfaces by Flow Microcalorimetry, *Carbon*,
      15, 311 (1977).

67.   V. J. Mimeault, D. W. McKee, 10th Conf. Carbon, Bethlehem,
      Paper No. FC. 25, 1971.

68.   J. Yamaki, The carbon Fiber/Acetal Copolymer Compositers:
      Adhesion Strength Between Fiber and Matrix for Discontinuous
      Fiber Composites II, *Kobunshi Ronbunshu*, *33*, 367 (1976).

69.   R. Yosomiya, T. Fujisawa, and M. Kobayashi, Graft Copolymeri-
      zation of Acrylonitril and Acrylonitril Styrene onto Carbon

Fiber Containing Unsaturated Groups, *Angew, Makromol. Chem.*, *118*, 183 (1983).

70. T. Fujisawa, A. Tanaka, and R. Yosomiya, Grafting of Styrene onto Carbon Fiber Having Pendent Thiol Groups, *Makromol. Chem.*, *183*, 2924 (1982).

71. R. Yosomiya and T. Fujisawa, Grafting of Styrene onto Carbon Fiber Having Perester Groups, *Polym. Bull.*, *13*, 7 (1985).

72. A. Tanaka, T. Fujisawa, and R. Yosomiya, Studies on the Graft Polymerization of Carbon Fiber, *J. Polym. Sci. Polym. Chem. Ed.*, *18*, 2267 (1980).

73. R. Yosomiya, T. Fujisawa, and K. Morimoto, Carbon Fiber Having Isocyanate Groups on the Surface, *Polym. Bull.*, *12*, 523 (1984).

74. M. Brie and C. L. Gressus, Grafting of Polymers on Carbon Fibers, *Fibre Sci. Technol.*, *6*, 47 (1973).

75. G. Riess and M. Bourdeaux, M. Bire, and G. Jouquet, Surface Treatment of Carbon Fibers with Alternating and Block Copolymers, *Proc. 2nd Carbonfiber Conf.*, *8*, 52 (1974).

76. R. V. Subramanian and J. J. Jakubowski, Electropolymerization on Graphite Fibers, *Polym. Eng. Sci.*, *18*, 590 (1978).

77. R. V. Subramanian, Electrochemical Polymerization and Deposition on Carbon Fibers, *Pure Appl. Chem.*, *52*, 1929 (1980).

78. R. Kondo and T. Hayashi, Interface Reaction of Composite Material, *Kagaku Sosetsu Jpn.*, *8*, 201 (1975).

79. I. Souma and K. Nohara, The Filler Effects of Xonotlite on Polyvinylchloride, *J. Rubber Soc. Jpn.*, *53*, 443 (1980).

80. K. Suzuki, Wollastnite, *Plast. Jpn.*, *31*, 78 (1980).

81. H. Ogino and S. Okazaki, The Surface Treatment of Fibrous Gypsum with Organic Acids and Bases, *Chem. Soc. Jpn.*, *1*, 20 (1980).

82. M. Okubo, S. Watatani, Y. Kusaka, and T. Matsumoto, Heterogeneous Polymerization of Methylmethacrylate in the Presence of Attapulgite in an Aqueous Medium, *J. Soc. Adhes. Jpn.*, *20*, 327 (1984).

83. K. Mitsuishi, S. Kodama, H. Kawasaki, and M. Tanaka, Effect of modified calcium carbonate on mechanical properties of Polypropylene/thermoplastic elastomers, *Kobunshi Ronbunshu*, *45*, 505 (1988).

84. M. Xanthos and R. T. Woodhams, Polymer Encapsulation of Colloidal Asbestos Fibrils, *J. Appl. Polym. Sci.*, *16*, 381 (1972).

85.   S. Chijiwa, Potassium Titanate Fiber Reinforced Plastics, *Kogyo Zairyo*, *29*, 44 (1981).

86.   Y. Yamaguchi, S. Sato, Y. Uchiyama, S. Ono, and M. Akiyama, Effects of Potassium Titanate Whisker and Coupling Agent on the Mechanical Strength of Polybutylene Terephthalate and Polyacetal Composites, *Bull. Kougakuin Univ. Jpn.*, *52*, 14 (1982).

87.   Y. Oyanagi, Y. Yamaguchi, K. Kubota, H. Itoi, and T. Narita, The Effects of Filled Potassium Titanate Whiskers on Some Physical and Mechanical Properties of Polyphenylene Sulfide Moldings, *Bull. Kougakuin Univ. Jpn.*, *56*, 31 (1984).

88.   S. Inubushi, T. Ikeda, and S. Tazuke, Excellent Flexural Properties of Aminimide-cured Epoxy Resin as a Matrix for Mica-dispersed Polymer Composites, *J. Mater. Sci.*, *23*, 1182 (1988).

89.   S. Inubushi, T. Ikeda, and S. Tazuke, Aminimide-cured Epoxy Resins as Surface Modifiers for Mica Flakes in Particle-reinforced Thermoplastics, *J. Mater. Sci.*, *23*, 535, (1988).

6

Surface Modification and Adhesion Improvement by the Blend Method

6.1 INTRODUCTION

The polymer blend method has been of particular interest where polymers with a specific molecular structure have migrated to the surface or the interface of the blend and have concentrated there. This method utilizes the microphase separation phenomena of graft polymers or block polymers blended in small amounts so that polar groups of the graft or block polymers are effectively concentrated on the surface or the interface of the blends.

This method enables the use of polymers of the same type as, or compatible with, the matrix polymer as the backbone polymer of the graft or block copolymers, so that composite materials that have uniform properties and are not separated into a matrix phase and a concentration phase in the vicinity of the surface or the interface can be obtained.

At present, the primary applications of this method are for the surface modification of polymers, that is, for rendering the polymer surface hydrophilic or hydrophobic. Few applications of the blend method for interface modification of composite materials have been found. However, it is believed that this method will be of great importance in such utilization fields.

Also, polymer-metal plate materials bonded strongly to each other were obtained by blending triazine thiols with polymers or by using triazine thiols as treating agents for metal plates, as described in Section 6.3. In this chapter, surface modifications and adhesion improvements using primarily the foregoing two methods are described.

6.2   SURFACE MODIFICATIONS BY POLYMER BLEND

6.2.1   Phase Separation Between Polymer—Polymer and Surface Adsorption of Graft or Block Copolymer

In many cases, use of a single polymer brings unsatisfactory results for the production of polymeric materials having the requisite performance and properties. For this reason, a plurality of polymers are often blended so as to obtain the desired properties. However, when mixed, the free energy of polymers does not take on a negative value, because the polymers are of large size, and changes in the entropy by mixing them does not contribute significantly to the free energy.

Except for combinations of polymers having an especially strong interaction (large negative enthalpy), it is difficult to blend different polymers and keep the blend in a uniform state for a long period of time. Generally, polymers are readily separated from each other (so-called macrophase separation). On the other hand, in block copolymers (polymeric constituents A and B are incorporated into the same chain) and graft copolymers (a backbone polymeric constituent A and a branch polymeric constituent B are chemically bonded), the same type of the constituent is inhibited from cohesion to another of the same type in the block or graft copolymer. As a result, no macrophase separation occurs, but a microphase-separated structure with fine domains is formed in the block or graft copolymer. Many useful reports have been published on the microphase separation mechanism [1—3] and structure [4—7] and are cited herein.

The characteristics of block and graft copolymers contribute to the amphiphilic property. In a solution, the block and graft copolymers form micelles, which leads to a domain structure in the solid state. Also, the copolymers are made to adsorb on the solution or solid surface, where the copolymers exhibit their characteristics.

Figure 6.1 shows a model of surface adsorption [46]. It is supposed that in an aqueous solution, the hydrophilic constituent of a hydrophobic-hydrophilic graft or block copolymer is extended in water, and the hydrophobic groups take the form of loops or tails adsorbing to the surface of the aqueous solution, partially exposed to the air [8,9].

Table 6.1 illustrates the surface adsorption of a styrene-siloxane block copolymer dissolved in an organic solvent [10]. The surface area of the siloxane chains is dominated by an area of compressed coils of the styrene chains formed in the organic solution. This is true of all block copolymers adsorbed to a solid surface.

In a block copolymer adsorbed to a solid surface through polar groups, the hydrophobic groups project into the organic solvent. The steric repulsion between the bulky hydrophobic groups stabilizes the dispersion system in the organic solvent [11], and this steric repulsion is dominated by an entropic factor [12], which is peculiar

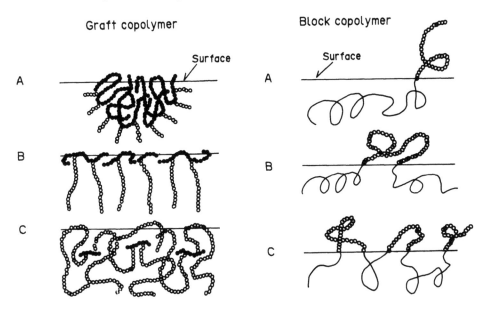

Figure 6.1   Schematic representation of surface segregation states
of graft and block copolymers with different graft and block archi-
tectures. (Ref. 46.)

to polymeric chains. This is one of the characteristics of polymer
dispersions and with increase in molecular weight, more stable dispersion systems can be obtained.

When styrene is polymerized in a hydrocarbon using a styrene-
siloxane block copolymer as a dispersant, the surface of the polystyrene particles is covered with siloxane chains, as reported previously. The surface area of siloxane chains on styrene particles
can be calculated. Results have revealed that these siloxane chains
are considerably elongated compared with random coil siloxane
chains [13].

It is possible to modify polymers by utilizing the surface adsorption of such block copolymers. For example, the surface of a film
formed by casting from a solution of dimethylsiloxane/bisphenol A
carbonate (BPAC) block copolymer with 1% polycarbonate was made
hydrophobic, as indicated by the wetting property (the contact angle
with ethylene glycol) in Table 6.2 [14].

Generally, in the case of a film surface contacting the air or,
for example, Teflon (fluororesin), the hydrophobic groups are arranged on the surface. On the other hand, when the film is in contact with a hydrophilic surface of glass or a similar substance, the
hydrophilic groups migrate to the film surface. Similarly, when the

Table 6.1 Surface Properties of Copolymers

Copolymer[a]	Liquid solution						Solid solution	
	Surface tension at 1% w/w (dyn/cm)	Concentration at start of plateau (% w/w)	Gibbs limiting area, A_L ($Å^2$/mol)	Polystyrene coil dimensions			Copolymer[a]	Critical surface tension of wetting (dyn/cm)
				Gibbs radius, r_a (Å)	Radius of gyration, $(\bar{s}^2)^{1/2}$ (Å)	$r_a/(\bar{s}^2)^{1/2}$		
Styrene	31.4						Polystyrene	32.7
Styrene + A	27.3	>10	670	10.3	18.1	0.57	Polystyrene + 1.0% A	28.3
Styrene + B	25.8	3.5	590	9.7	15.7	0.62	Polystyrene + 1.0% B	27.5
Styrene + C	25.3	2.0	550	9.4	12.8	0.73	Polystyrene + 1.0% C	22.0
Styrene + D	23.5	0.5	280	6.7	9.1	0.74	Polystyrene + 1.0% D	22.0
Styrene + E	25.5	2.0	470	8.6	11.1	0.77		
Styrene + F	25.8	2.0	340	7.4	9.1	0.81		

[a]Polydimethylsiloxane % (calculated from monomer consumption). A: 20% ($\bar{M}n$ = 10,000), B: 37% ($\bar{M}n$ = 9,500), C: 59% ($\bar{M}n$ = 9,700), D: 78% ($\bar{M}n$ = 9,100), E: 58% ($\bar{M}n$ = 7,300), F: 100% ($\bar{M}n$ = 4,800).
Source: Ref. 10.

Table 6.2   Contact Angle to Ethylene Glycol
of Polycarbonate Film Cast Against Various
Substrates

Cast substrate	θ (deg)
Poly(BPAC)	
Air side	56
Against Teflon	67
Against clean glass	53
Poly(BPAC) + 1%(DMS)$_{40}$−(BPAC)$_{6.6}$	
Air side	89
Against Teflon	89
Against clean glass	59
Same, Annealed at 140°C × 18 h	91
Same, Rinsed with solvents	90
Poly(DMS)	92
Block copolymers (DMS)$_{20}$−(BPAC)$_{3.9}$	89

Source: Ref. 14.

film is heat-pressed, the film surface becomes hydrophobic or hydrophilic, depending on the type of contact surface.

Figure 6.2 shows the measurement results of the contact angles
of films [10] formed by pressing from polystyrene added to 1%
styrene-siloxane block copolymer.   Evidently, the elongated siloxane
chains indicate that the polystyrene surface is covered with siloxane
chains.   Such effects are inherent in block copolymers that have
relatively long chains and particular compositions.   Also, in the case
of siloxane-nylon block copolymers, such surface modification effects
of the siloxane block copolymers have been confirmed [15].

According to a study by Gains and co-workers [16] regarding
elucidating these phenomena, polystyrene was added with a small
amount of styrene-siloxane block copolymer, and the surface tensions were measured by the drop weight method, as shown in Figure
6.3.   The behavior of added block copolymer was analyzed from the
reduction curve of the surface tension.   A mechanism was suggested
whereby the spherical domains of the polysiloxane were diffused
through the polystyrene and surface-adsorbed.   Adsorption equilibrium was set up from the beginning of the addition at addition
amounts above 0.2%.   The adsorption amount increased with elapsed
heating time at addition amounts below 0.2%.

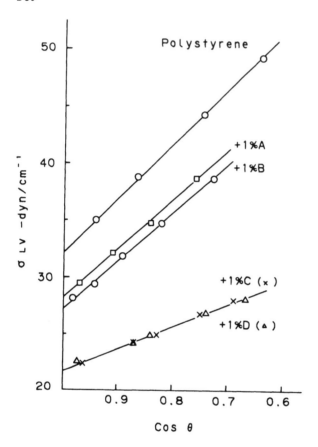

Figure 6.2   Contact angle plots for the polystyrene disks containing 1.0% each of four block polymers. A, B, C, D: Table 6.1. (Ref. 10.)

Such effects were observed in the addition of styrene-tetrahydro-furan block copolymers. The block copolymers are phase-separated and present a typical "islands in the sea" structure. However, in relatively low amounts of the tetrahydrofuran constituent, the spheri-cal domains of the tetrahydrofuran are diffused to the surface and concentrate there [17]. The surface is ready to be wetted with water, since the tetrahydrofuran segments are hydrophilic. Figure 6.4 shows the test results [18a]. By utilizing these results, the surface of polystyrenic substances can be rendered hydrophilic by the addi-tion of block copolymers at amounts below 1% [18b].

6.2.2   Surface Modification by Macromers and by Macromer Graft Copolymers

Surface modification can be achieved by blending a small amount of the block or graft copolymers as described above. For such migration and concentration of block or graft copolymers to be controlled efficiently, it is essential that the molecular structure of the block or graft copolymers themselves be adequately controlled.

The macromer method is exemplified by an effective synthetic process developed expressly for this purpose. Macromer a registered

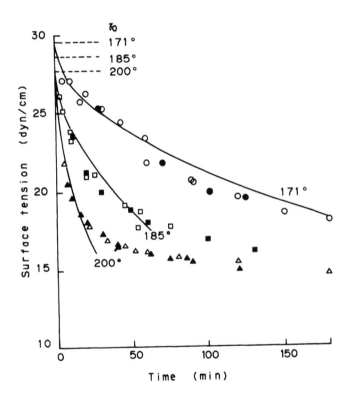

Figure 6.3   Surface tension as a function of time for a blend of polystyrene + 0.05 wt % poly(styrene$_{75}$-co-dimethylsiloxane$_{77}$). Experimental points: o, 171°; □, 185°; △, 200°; shading in symbols indicates different runs. Curves are calculated for diffusion-controlled adsorption; see the text. γ_0 is the surface tension of pure polystyrene, M = 9290, at the indicated temperature. (Ref. 16a.)

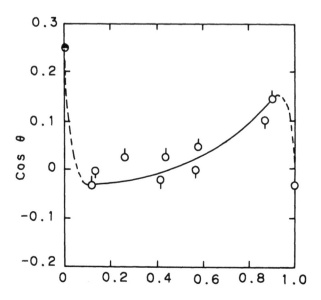

Styrene mole fraction in block copolymer

Figure 6.4   Relationships between composition and wettability of ST
and STS block copolymers.   ○, polystyrene; ●, polytetrahydrofuran;
ᵟ, ST; ♀, STS; θ, contact angle.   (Ref. 17, 18a.)

trade name based on the term "macromolecular weight monomer," is
a method that enables polymerization. Many studies concerning the
Macromer method have been undertaken. Applications of the Macromer
concept for the synthesis of graft copolymers have increased rapidly.

In the Macromer method for the synthesis of graft copolymers,
the parts of the graft copolymers that form the branches have previously been synthesized. Accordingly, differences between the
conventional and Macromer methods, or the advantages of the Macromer method, are based on the following considerations:

1. A desired function can be incorporated into the branch.
2. The branches are introduced into the copolymer structure
    by the copolymerization method. Accordingly, the length
    and number of branches can easily be controlled by utilizing
    the known or measured reactivity of the monomer.
3. There is less possibility that a homopolymer of branch units
    exists in the graft copolymers, because of the low homopoly-
    merizability of the macromer.

By the popular Macromer synthetic method, carboxylic or alcohol-ended prepolymers are synthesized by radical polymerization utilizing a chain-transfer agent, and then double bonds are introduced by conversion of the end groups. The combination of acrylic monomers and mercaptan chain-transfer agents is particularly effective for synthesizing macromers with a molecular weight of 2000 to 8000 [19,20]. The following reaction formulas show typical macromer syntheses.

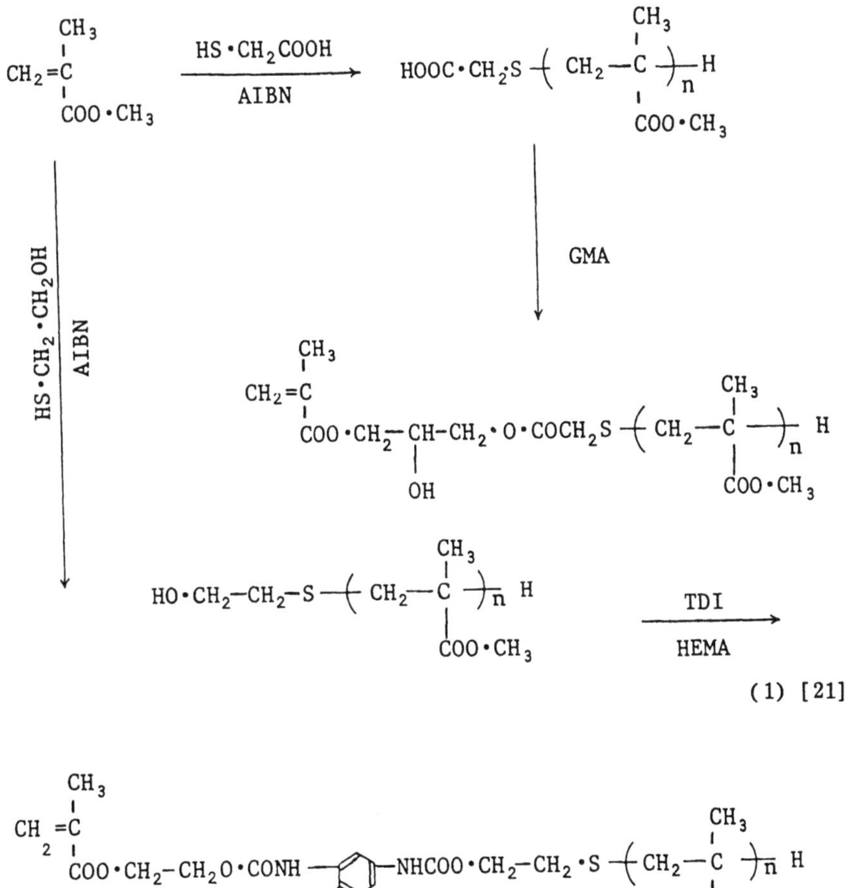

$$CH_2=\underset{\underset{COO \cdot CH_3}{|}}{\overset{\overset{CH_3}{|}}{C}} \quad \xrightarrow[\text{AIBN}]{\text{HS} \cdot CH_2COOH} \quad HOOC \cdot CH_2S \left(CH_2 - \underset{\underset{COO \cdot CH_3}{|}}{\overset{\overset{CH_3}{|}}{C}} \right)_n H$$

$$\downarrow \text{GMA}$$

$$CH_2 = \underset{COO \cdot CH_2-CH-CH_2 \cdot O \cdot COCH_2 S}{\overset{CH_3}{|}}{C} \underset{\underset{OH}{|}}{} \left(CH_2 - \underset{\underset{COO \cdot CH_3}{|}}{\overset{\overset{CH_3}{|}}{C}} \right)_n H$$

$$HS \cdot CH_2 \cdot CH_2OH \quad | \quad AIBN \downarrow$$

$$HO \cdot CH_2 - CH_2 - S \left(CH_2 - \underset{\underset{COO \cdot CH_3}{|}}{\overset{\overset{CH_3}{|}}{C}} \right)_n H \quad \xrightarrow[\text{HEMA}]{\text{TDI}}$$

(1) [21]

$$CH_2 = \underset{COO \cdot CH_2 - CH_2O \cdot CONH - \bigcirc - NHCOO \cdot CH_2 - CH_2 \cdot S}{\overset{CH_3}{|}}{C} \underset{CH_3}{} \left(CH_2 - \underset{\underset{COO \cdot CH_3}{|}}{\overset{\overset{CH_3}{|}}{C}} \right)_n H$$

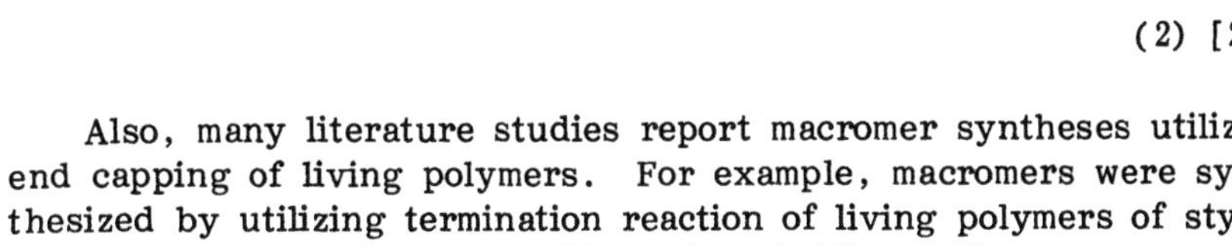

(2) [22]

Also, many literature studies report macromer syntheses utilizing end capping of living polymers. For example, macromers were synthesized by utilizing termination reaction of living polymers of styrene, isoprene, and so on, with various halides [23].

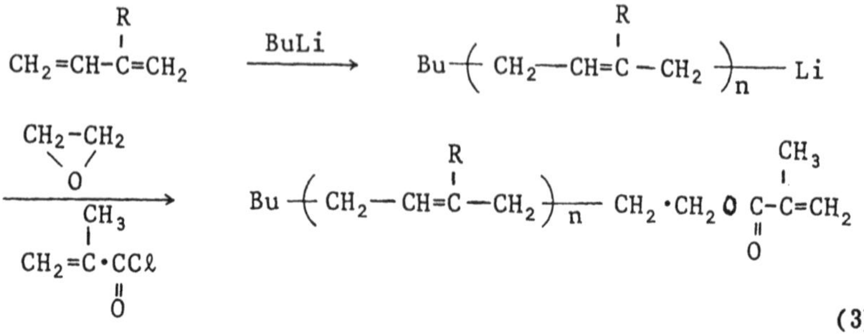

(3)

It was difficult to living-polymerize polar vinyl monomers and to introduce ended double bonds capable of being polymerized quantitatively. Recently, the following new polymerization technique has been developed. The synthesis of macromers having an introduction ratio of functional groups of about 100% has been realized [24,25].

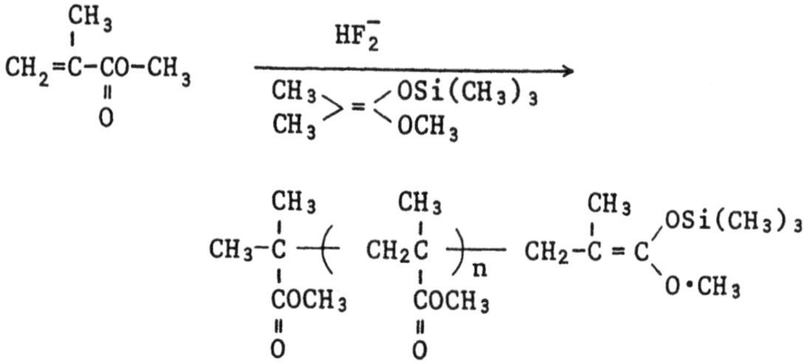

(4)

The reactivity of the ended double bond is often a problem for applications of the foregoing macromers. The data reported previously indicate that the reactivity of macromers is as high as that of the corresponding low-molecular-weight models [26–29]. Figure 6.5 shows a composition curve of copolymers of poly(methyl methacrylate) with stearyl methacrylate [20]. In general, it is thought that the macromer exhibits almost the same reactivity as that of the corresponding low-molecular-weight monomer, at least in the copolymerization reaction.

The above-mentioned graft copolymers have interfacial activity. When blended with other polymeric solids, the graft copolymers are deposited and adsorbed to the surface of the solids. As one utilization of the phenomenon noted above, the surface of polymeric base materials can be modified and rendered a function by using a tailored graft copolymer.

Figure 6.6 shows the degree of surface modification of poly(methylmethacrylate) by measurement of the contact angle [29]. The surface

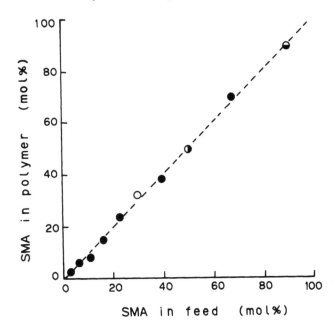

Figure 6.5   Copolymer composition data for copolymerization of SMA with •, MMA, 9–22 wt % conversions; ○, PMMA macromonomer ($\bar{M}n$ = 1650), 39 wt % conversion; ○, PMMA macromonomer ($\bar{M}n$ = 4080), 61 wt % conversion; ○, macromonomer ($\bar{M}n$ = 4080), 35 wt % conversion. (Ref. 20.)

of this poly(methyl methacrylate) was modified using a graft copolymer of which the backbone of which is composed of poly(fluoroalkyl acrylate) (which has a water-repelling function) and the branch parts of poly(methyl methacrylate) macromer. That is, a small amount of the graft copolymer was added to poly(methyl methacrylate), dissolved in benzene, and formed into a film. The contact angles between water and the film surface, which is in contact the air and the glass, were measured. The results have revealed that the air-side surface of the poly(methyl methacrylate) film is modified nearly to the water repellency of fluorocarbons by the addition of 0.2% graft copolymer. On the other hand, random copolymers with the same composition as the block copolymers have relatively less effect on the surface modification [30,31].

Such surface modifications can also be achieved by using graft copolymers that contain hydrophobic segments as the branch. For example, small amounts of graft copolymers containing poly(dimethyl

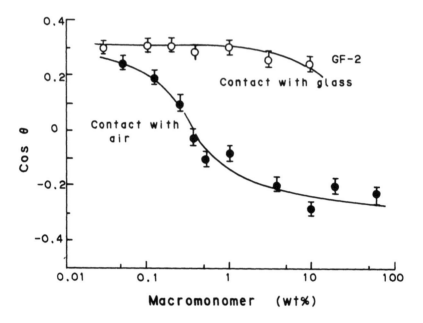

Figure 6.6 Surface modification of poly(methyl methacrylate) by poly(fluoroalkyl acrylate)-methyl methacrylate macromonomer. (Ref. 29.)

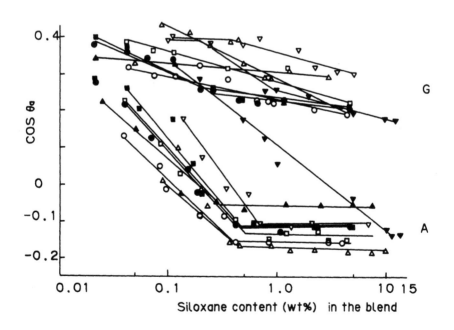

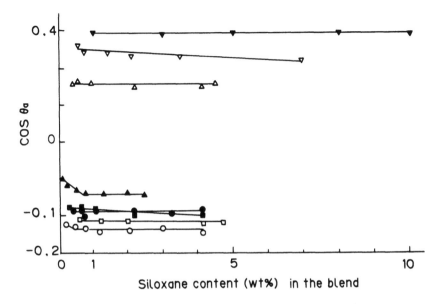

Figure 6.8   Contact angle of water droplet at 20°C on the air-side surface of various silicone polymer-PMMA blend films after treatment with *n*-hexane. (For an explanation of the symbols, and for contact angle values before *n*-hexane treatment, see Figure 6.7.) (Ref. 31.)

siloxane) macromer as the branch were added to poly(methyl methacrylate) and formed into a film on a glass plate.

Figure 6.7 shows values for contact angles of both the air-contact side and the glass-contact side film surface [31]. The measurement results reveal that the surface of poly(methacrylate) is rendered as hydrophobic as poly(dimethyl siloxane) for a graft polymer addition amount of about 1%. The durability of such surface modifications was examined by surface treatment with *n*-hexane, which is a good solvent for poly(dimethyl siloxane). No reductions in the surface hydrophobic property of the film were found.

Figure 6.7   Contact angle of water droplet at 20°C for various silicone polymer-PMMA binary blends (A, air-side surface; G, glass-side surface). The symbols represent the following silicone polymers: ○, GM-211; ●, GM-213; □, GM-411; ■, GM-413; ▼, PDMS; ▽, PMTS; ▲, MTS-45; △ MTS-25. (Ref. 31.)

In the case of systems to which random copolymer or homopolymer was added, the reduction in contact angle was remarkable. Figure 6.8 shows the test results. It is generally understood that such
differences in durability are caused by the functions of the poly-
(methyl methacrylate) backbone constituent, which dissolved into
the base poly(methyl methacrylate) as an anchor.

Only recently have attempts to modify solid surfaces by utilizing
the surface activities of block and graft copolymers been made.
There are few applications in the composite material fields. Typically, the substitutional use of a polar group containing block and
graft copolymers for silane coupling agents, improvement in the interfacial adhesion of matrix resins by addition of these block and
graft copolymers, then deposition of the copolymers to the resin surface, and so on, are suggested.

6.3 ADHESION IMPROVEMENT BY TRIAZINE THIOLS BLEND

6.3.1 Reactions of Triazine Thiols

Many studies on cross-linking adhesion between different materials
by using triazine thiols have recently been made [32]. With regard
to obtaining high-quality composite materials, blending of triazine
thiols with filler-matrix systems is of wide interest as one method
of enhancing interaction between the filler and the matrix. Table
6.3 lists typical triazine thiols (the abbreviations are designated in
parentheses.) The various reactions of triazine thiols previously
reported are illustrated below.

$$\text{TSS} + \underset{\text{halogenated rubber}}{\overline{\quad\quad\underset{x}{|}\quad\quad}} \xrightarrow{\text{MgO}} \overline{\quad\underset{\text{TSS}}{|}\quad} + \text{MgX}_2$$

$$\xrightarrow{\text{MgO}} \quad \begin{matrix} \overline{\quad\quad} \\ | \\ S \\ T \\ S \\ | \\ \underline{\quad\quad} \end{matrix} \qquad (5)$$

In coexistence with metal oxides such as MgO, triazine thiols function as a cross-linking agent for halogen-containing polymers [33].

$$\text{TSS} + \left(\!\!\text{CH}_2\text{-CH}_2\!\!\right)\!\!-\!\!\underset{\text{PE}}{} + \text{ROOR} \longrightarrow \begin{matrix} \sim\!\!\sim\text{PE} \\ | \\ S \\ T \\ S \\ | \\ \sim\!\!\sim\text{PE} \end{matrix} + 2\text{ROH} \qquad (6)$$

In coexistence with peroxides, triazine thiols function as a cross-linking agent for polyethylene, butyl rubber, and so on.

$$TSS + CH_2-CH \cdot CH_2 \cdot R \longrightarrow \begin{array}{l} S - CH_2 \cdot CH(OH) \cdot CH_2R \\ T \\ S - CH_2 \cdot CH(OH) \cdot CH_2R \end{array} \qquad (7)$$

$$TSS + -OCOCH=CHCOO- \longrightarrow \begin{array}{l} -OCO \cdot CH \cdot CH_2 \cdot COO- \\ \quad\quad | \\ \quad\quad S \\ \quad\quad T \\ \quad\quad S \\ \quad\quad | \\ -OCO \cdot CH \cdot CH_2COO- \end{array} \qquad (8)$$

$$TSS + MO(M) \longrightarrow TSS-M-SST + H_2O \qquad (9)$$

Reaction formulas (7) and (8) express the functions of triazine thiols as a curing or cross-linking agent for epoxy compounds and unsaturated polyesters. Formula (9) expresses the reaction of triazine thiols with metal oxides [34,35]. Triazine thiols have a high degree of reactivity with many polymeric, inorganic, and metallic materials which implies that triazine thiols are useful for the surface modification of composite materials.

Table 6.3 Properties of Triazine Thiols

	$-R$	M	pK_a
F(TT)	$-SH$	H(Na)	6.5
AF(AN)	$-NHC_6H_5$	H(Na)	5.5
DB(DBN)	$-N(C_4H_9)_2$	H(Na)	4.1
DA	$-N(CH_2CH=CH_2)_2$	H	4.2
OL	$-NHC_{18}H_{35}$	Na	–

Source: Ref. 32.

6.3.2 Modification of Polymers by Triazine Thiols for Adhesion

The bonding of polymers to other polymers is dependent on the solubility parameters and reactivities of the polymers used, whether or not bonding is possible. From the viewpoint of solubility parameters and reactivity, polymers are grouped as follows: (1) the solubility parameters and reactivities are both different; (2) either the solubility parameters or the reactivities are different; or (3) the solubility parameters and reactivities are about the same.

In case 1, where the polymers are bonded with most difficulty, typically in bonding of poly(vinyl chloride) (PVC) to an ethylene-propylene-diene terpolymer (EPDM), PVC is added with an appropriate plasticizer to adjust the apparent solubility parameter to close to that of EPDM, and also with triazine thiols, so that a strong adhesion between PVC and EPDM can be obtained. Table 6.4 illustrates the addition effects of the triazine thiols used for adhesion of PVC to various rubbery polymers [36]. The added chloranil functions as an accelerator of the reaction between the thiol groups of the triazine thiols and the EPDM molecules [37].

In bonding of PVC to polyethylene (PE), a PVC sheet containing a triazine thiol and MgO was overlaid on a polyethylene sheet containing a triazine thiol and a peroxide, and then hot-pressed, so that the PVC-PE sheets were strongly bonded, which is expressed by reaction formula (10). Figure 6.9 shows the test results [38,39].

$$(10)$$

Generally, fluororubbers are difficultly bonded polymeric materials. However, fluororubbers containing a triazine thiol and a tetraonium salt in an appropriate combination can be bonded sufficiently well to

Table 6.4   Vulcanizing Adhesion of PVC to Rubber[a]

Compound[b]	Triazine trithiol in PVC compound[c] (phr)	Rubber	sp Value	Chloranil	Peel strength (kN/m)
1	–	EPDM	7.9	–	0.3
	3	EPDM		–	2.0
	–	EPDM		1	0.3
	3	EPDM		1	3.0
2	–	IR	7.9	–	0.5
	3	IR		–	1.2
	–	IR		1	0.8
	3	IR		1	2.4
3	–	BR	8.4	–	0.3
	3	BR		–	4.8
	–	BR		1	0.6
	3	BR		1	5.7
4	–	SBR	8.6	–	0.6
	3	SBR		–	4.2
	–	SBR		1	0.8
	3	SBR		1	6.6
5	–	CIR	8.0	–	0.8
	3	CIR		–	3.4
6	–	CO	9.4	–	0.6
	3	CO		–	9.6
7	–	CR	9.4	–	1.2
	3	CR		1	10.4
8	–	NBR	9.6	–	4.0
	3	NBR		–	7.2
	–	NBR		1	4.7
	3	NBR		1	14.5

[a] sp value of PVC; 9.6; press conditions: 170°C × 30 min.

[b] Rubber compound:
1. EPDM 100 parts, SRF black 50 phr, oil 30 phr, MBTS 2 phr, sulfur 0.5 phr, ZnO 5 phr, St 1 phr, chloranil 0 or 1 phr.
2. IR 100 parts, SRF black 50 phr, MBTS 2 phr, sulfur 0.5 phr, ZnO 5 phr, St 1 phr, chloranil 0 or 1 phr.
3. BR 100 parts, SRF black 50 phr, MBTS 2 phr, sulfur 0.5 phr, ZnO 5 phr, St 1 phr, chloranil 0 or 1 phr.

Table 6.4   (Continued)

4. SBR 100 parts, SRF black 50 phr, MBTS 2 phr, sulfur 0.5 phr, ZnO 5 phr, St 1 phr, chloranil 0 or 1 phr.
5. CIR 100 parts, SRF black 50 phr, 6-dibutylamino-1,3,5-triazine-2,4-dithiol 3 phr, MgO 5 phr, St 1 phr.
6. CO 100 parts, SRF black 50 phr, triazine trithiol 2 phr, MgO 3 phr, $CaCO_3$ 5 phr.
7. CR 100 parts, SRF black 50 phr, triazine trithiol 2 phr, MgO 5 phr, ZnO 5 phr, St 1 phr.
8. NBR 100 parts, SRF black 50 phr, oil 10 phr, MBTS 2 phr, sulfur 0.5 phr, ZnO 5 phr, St 1 phr.

^{c}PVC compound: PVC 100 parts, DOP 50 phr, triazine trithiol 0 or 3 phr, MgO 5 phr, Sta. 1 phr.
Source: Ref. 36.

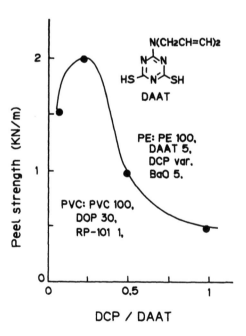

Figure 6.9   Cross-linking adhesion between PVC and PE.   (Ref. 36.)

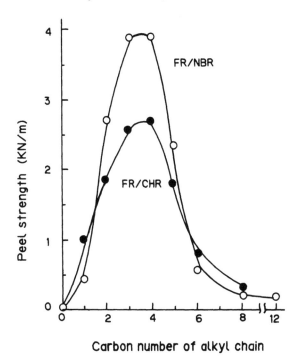

Figure 6.10   Influence of carbon number of $(C_nH_{2n} + 1)_4$ NBr on the peel strength of adherends. ○, cross-linking adherends of FR/NBR; •, cocross-linking adherends of FR/CHR. Adhesion conditions: 160°C, 30 min under 5 MPa and postcure 180°C for 20 min. (Ref. 41.)

polar rubbers such as butadiene-nitrile copolymeric rubbers (NBR), epichlorohydrin rubbers, chlorinated rubbers, chloroprene rubbers (CR), and the like. Here the triazine thiol functions as a cross-linking agent, and the tetraonium salt acts as a catalyst for activating the thiol groups [40,41]. Also, the structure of the ammonium salt exerts a strong influence on the peeling resistance of the bonded sheets.

Figure 6.10 shows the effects of alkyl groups of tetraalkyl ammonium bromide with different carbon numbers on the adhesion of a fluororubber to NBR, a chlorohydrin-ethyleneoxide copolymeric rubber (CHR). The results support the suggestion that the basicity and migration capability of the ammonium salt are related to the reaction between the fluororubber and polymeric rubbers at the interface. As shown in Table 6.5, PVC-fiber materials with high bonding

Table 6.5   Effect of Fabrics[a] on Peel Strength
in Adhesion of PVC[b] Sheets to the Fabrics

Fabrics	Denier	Peel strength[c] (kg/3 cm)
Nylon 6,6	840	2.1 (0.2)
Nylon 6	420	3.5 (0.4)
	110	3.1 (0.4)
Polyester	1000	1.2 (0.1)
	100	2.0 (0.3)
	75	1.3 (0.2)

[a]Epoxy treatment: Epoxy G-100 5 g, methanol
100 ml, 20°C × 5 min, heat treatment for 30
min at 165°C. [b]PVC plastisol: Zeon 121 100
parts, DOP 60 phr, DB 3 phr, MgO 5 phr.
Adhesion condition: 180°C × 20 min.

[c](−) Blank values in parentheses.
Source: Ref. 45.

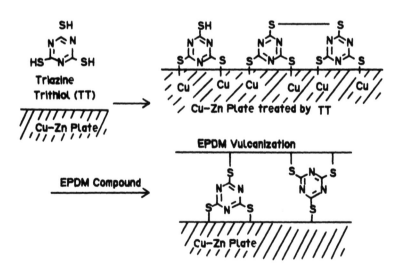

Figure 6.11   Schematic diagram of cross-linking model.   (Ref. 36.)

Table 6.6   Effects of Triazine Thiols and Solvents on the Peel Strength in the Vulcanizing Adhesion of EPDM to Brass Plates Treated with Triazine Thiol Solutions

Triazine thiol	Concentration (%)	Treatment conditions			Peel strength (kN/m)	
		Solvent	Temperature (°C)	Time (min)	PSah[a]	PSbh[b]
TT	0.03	PEG[c]	130	15	3.0	7.2
TT	0.3	PEG[c]	130	15	4.1	9.4
TT	1.0	PEG[c]	130	15	4.1	8.8
AN—Na	0.3	H_2O	80	5	4.0	8.9
DB—Na	0.3	H_2O	80	5	2.6	7.0
TT—Na	0.3	H_2O	80	5	2.0	2.1
TT	0.3	MeOH	60	15	3.4	6.5
TT	0.3	Solvso[d]	130	15	2.5	8.9
TT—Na	0.3	MeOH	60	15	3.2	8.0

[a]Peel strength before heat treatment.
[b]Peel strength after heat treatment.
[c]Polyethylene glycol (MW = 300).
[d]High-olefin oil (bp, 150–240°C).
Source: Ref. 42.

Table 6.7   Cross-Linked Adhesion of Metal and Rubber Containing
Triazine Thiols

Metal/rubber	Compound (phr)	°C × min	Peel strength (kN/m)
Cu/CHR	DB 1, AF 1, MgO 5, FEF 40	160 × 40	10.6
Cu/CHC	F 0.5, AF 1, CaCO$_3$ 10, FEF 40	170 × 30	12.6
Cu/SBR	F 3, MBTS 3, S 0.5, St 1, ZnO 5	160 × 20	17.0
Cu—Ni/SBR	F 3, MBTS 3, S 0.5, St 1, ZnO 5	160 × 20	8.5
Cu—Sn/SBR	F 3, MBTS 3, S 0.5, St 1, ZnO 5	160 × 30	12.7
Cu/NBR	F 3, MBTS 3, S 0.5, St 1, ZnO 5	170 × 30	12.0
Cu/EDPM	F 3, MBTS 3, S 0.5, St 1, ZnO 5	180 × 30	10.6
Cu/CR	F 3, MBTS 3, S 0.5, St 1, ZnO 5	170 × 20	7.6
Cu/PE	DCP 3	160 × 30	9.0
Cu/PVC	F 3, PEG 5, MgO 3, RP 101 DOP 30, FEF 10	180 × 20	9.8

Source: Ref. 32.

strengths were obtained by heat-pressing fibers of fabrics treated
with an epoxy compound against PVC containing a triazine thiol [45].

6.3.3   Adhesion of Polymers to Metals
Utilizing Triazine Thiols

Triazine thiols are able to react with metals and metal oxides and
form specific surface films as mentioned above [formula (9)].   When
a metal material was dipped into a triazine thiol solution, an organic
film dozens to several hundred of micrometers thick can be formed.
This film has characteristics such as high resistance to heat and
water, and inclusion of functional groups such as thiol, disulfide,

mercaptide groups, and so on, which are active for the polymer [42]. Accordingly, when a polymeric sheet is overlaid on the organic film and heated, the polymer molecules are diffused into the film layer and react with the active functional groups to form a strong adhesion layer.

Figure 6.11 shows a model of the adhesion mechanism [44]. Table 6.6 lists the effects of treatment condition on the adhesion between a triazine thiol-treated brass board and EPDM. The peeling strength is considerably enhanced by the heat treatment. Such improvements in adhesion between a brass plate and a polymer can be obtained without triazine thiol treatment of the brass plate simply by adding a triazine thiol to a polymer and bonding the polymer to an untreated brass plate [43,44]. Table 6.7 shows the test results. These results support the mechanism through which the triazine thiol molecules contained in the polymer are diffused to the interface between the polymer and the brass sheet, and react with the brass surface to form such a film.

REFERENCES

1. D. J. Meier, Theory of Block Copolymers: I. Domain Formation in A—B Block Copolymers, *J. Polym. Sci.*, *C26*, 81 (1969).

2. E. Helfand and Z. R. Wasserman, Block Copolymer Theory: 4. Narrow Interphase Approximation, *Macromolecules*, *9*, 879 (1976).

3. L. Leibler, Theory of Microphase Separation in Block Copolymers, *Macromolecules*, *13*, 1602 (1980).

4. T. Ono, H. Minamiguchi, T. Soen, and H. Kawai, Domain Structure and Viscoelastic Properties of Graft Copolymer, *Kolloid. Z. Z. Polym.*, *250*, 394 (1972).

5. H. Hasegawa and T. Hashimoto, Morphology of Block Polymers Near a Free Surface, *Macromolecules*, *18*, 589 (1985).

6. T. Hashimoto, Y. Tsukahara, K. Tachi, and H. Kawai, Domain Boundary Mixing and Mixing in Domain" Effects on Microdomain Morphology and Linear Dynamic Mechanical Response, *Macromolecules*, *16*, 648 (1983).

7. Y. Tsukahara, N. Nakamura, T. Hashimoto, and H. Kawai, Structure and Properties of Tapered Block Polymers of Styrene and Isoprene, *Polym. J.*, *12*, 455 (1980).

8. A. G. Kanellopoulos and M. J. Owen, The Adsorption of Polydimethylsiloxane Polyether ABA Block Copolymers at the

Water/Air and Water/Silicone Fluid Interface, *J. Colloid Interface Sci.*, *35*, 120 (1971).

9. M. Kawaguchi and A. Takahashi, Ellipsometric Study of the Adsorption of Comb-Branched Polystyrene onto a Metal Surface, *J. Polym. Sci. Polym. Phys. Ed.*, *18*, 943 (1980).

10. M. J. Owen and T. C. Kendrick, Surface Activity of Polystyrene-Polysiloxane-Polystyrene ABA Block Copolymers, *Macromolecules*, *3*, 458 (1970).

11. D. H. Napper, Flocculation Studies of Sterically Stabilized Dispersions, *J. Colloid Interface Sci.*, *32*, 106 (1970).

12. J. V. Dawkins and G. Taylor, Nonaqueous Polymethylmethacrylate Dispersion, Radical Dispersion Polymerization in the Presence of AB Block Copolymers of Polystyrene and Polydimethylsiloxane, *Polymer*, *20*, 599 (1979).

13. J. V. Dawkins and G. Taylor, Micelle Formation by AB Block Copolymers of Polystyrene and Polydimethylsiloxane in *n*-Alkans, *Makromol. Chem.*, *180*, 1737 (1979).

14. D. G. LeGrand and G. L. Gaines, Jr., Surface Activity of Block Copolymers of Dimethylsiloxane and Bisphenol-A Carbonate in Polycarbonate, *Polym. Prepr. Am. Chem. Soc. Div. Polym. Chem.*, *11*, 442 (1970).

15. M. J. Owen and J. Thompson, Siloxane Modification of Polyamides, *Br. Polym. J.*, *4*, 297 (1972).

16a. G. L. Gains, Jr., and G. W. Bender, Surface Concentration of a Styrene-Dimethylsiloxane Block Copolymer in Mixtures with Polystyrene, *Macromolecules*, *5*, 82 (1972).

16b. G. L. Gains, Jr., and G. W. Bender, Surface Studies on Multicomponent Polymer Systems by X-Ray Photoelectron Spectroscopy, *Macromolecules*, *12*, 1011 (1979).

17. A. Takahashi, H. Wakabayashi, K. Honda and T. Kato, Wettability and Composition of Styrene-Tetrahydrofuran Block Copolymers, *Kobunshi Ronbunshu*, *35*, 269 (1978).

18a. Y. Yamashita, Surface Properties of Styrene-Tetrahydrofuran Block Copolymers, *J. Macromol. Sci. Chem. Ed.*, *A13*, 401 (1979).

18b. A. Takahashi and Y. Yamashita, Morphology, Crystallization, and Surface Properties of Styrene-Tetrahydrofuran Block Polymers, *Adv. Chem.*, *142*, 267 (1975).

19. K. K. Roy, D. Pramanick, and S. R. Palit, Application of Dye Techniques in the Study of Chain-Transfer Properties of Thiols, *Makromol. Chem.*, *153*, 71 (1972).

20. K. Ito, N. Usami, and Y. Yamashita, Syntheses of Methyl Methacrylate-Stearyl Methacrylate Graft Copolymers and Characterization by Inverse Gas Chromatography, *Macromolecules*, *13*, 216 (1980).

21. K. E. J. Barrett, *Dispersion Polymerization in Organic Media*, John Wiley & Sons Ltd., Chichester, West Sussex, England, 1975.

22. B. W. Jackson, U.S. Pat. 3,689,593 (1972).

23. G. O. Shulz and R. Milkovich, Graft Polymers with Macromonomers: I. Synthesis from Methacrylate-Terminated Polystyrene, *J. Appl. Polym. Sci.*, *27*, 4773 (1982).

24. O. W. Webster, W. R. Hertler, D. Y. Sogah, W. B. Farnham, and T. V. RajanBabu, Group-Transfer Polymerization: I. A New Concept for Addition Polymerization with Organosilicon Initiator, *J. Am. Chem. Soc.*, *105*, 5706 (1983).

25. D. Y. Sogah and O. W. Webster, Telechelic Polymers by Group Transfer Polymerization, *J. Polym. Sci. Polym. Lett. Ed.*, *21*, 927 (1983).

26. Y. Yamashita, Synthesis of Anphiphilic Graftcopolymers from Polystyrene Macromonomer, *Polym. J.*, *14*, 255 (1982).

27. M. Maeda, Y. Nitadori, and T. Tsuruta, Synthesis of New Monomers Having a Primary Amino Group by Lithium Alkylamide Catalyzed Addition Reaction of *N*-Alkylethylenediamines with 1,4-Divinylbenzene, *Makromol. Chem.*, *181*, 2251 (1980).

28. A. Revillon and T. Hamaide, Macromer Copolymerization Reactivity Ratio Determined by GPC Analysis, *Polym. Bull.*, *6*, 235 (1982).

29. Y. Yamashita, Y. Tsukahara, K. Ito, M. Okada, and Y. Tajima, Synthesis and Application of Fluorine Containing Graftcopolymers, *Polym. Bull.*, *5*, 335 (1981).

30. Y. Yamashita and Y. Tsukahara, Modification of Polymer by Tailoerd Graft Copolymers, *Polym. Sci. Technol.*, *21*, 131 (1983).

31. Y. Kawakami, R. A. Murthy, and Y. Yamashita, Surface Active Properties of Silicone Containing Polymers, *Polym. Bull.*, *10*, 368 (1983).

32. K. Mori and Y. Nakamura, Stabilization of Interface of Different Materials, *Surf. Jpn.*, *23*, 709 (1985).

33.   K. Mori and Y. Nakamura, Crosslinking of Halogen-Containing
       Rubbers with Triazine Dithiols, *Rubber Chem. Technol.*, *57*,
       34 (1984).

34.   K. Mori and Y. Nakamura, Action of Triazine Thiols and Their
       Metal Salts to Peroxide, *J. Appl. Polym. Sci.*, *26*, 691 (1983).

35.   K. Mori, Functionality of Metal Surface, *Denki Kagaku*, *54*, 96
       (1986).

36.   K. Mori and Y. Nakamura, Crosslinking Adhesion Between Dif-
       ferent Materials, *J. Adhes. Soc. Jpn.*, *19*, 382 (1983).

37.   K. Mori and Y. Nakamura, Adhesion of Rubbers to Polyvinyl-
       chloride During Vulcanization, *Plast. Rubber Process. Appl.*,
       3, 17 (1983).

38.   Y. Nakamura, K. Mori, and H. Nishina, Improvement of Ad-
       hesion Properties and Compatibility Between Polyethylene and
       Polyvinylchloride by Crosslinking, *J. Adhes. Soc. Jpn.*, *21*,
       95 (1985).

39.   Y. Nakamura, K. Mori, Y. Yoshida, and K. Tamura, Improve-
       ment of Adhesion and Compatibility between Polyethylene and
       Polyvinylchloride by Crosslinking, *Kobunshi Ronbunshu*, *41*,
       531 (1984).

40.   Y. Nakamura, K. Mori, and K. Wada, Crosslinking Adhesion
       of Fluoro Rubber to Nithile of Epichlorohydrin Rubber, *J. Soc.
       Rubber Ind. Jpn.*, *57*, 561 (1984).

41.   Y. Nakamura, K. Mori, and K. Wada, Improvement of Some P
       Properties of Fluorine-Containing Crosslinked Blend Rubber
       by Crosslinking, *Kobunshi Ronbunshu*, *41*, 539 (1984).

42.   K. Mori, Y. Nakamura, M. Shida, and I. Nishiwaki, Vulcanizing
       Adhesion of EPDM on Surface Treated Brass Plate by Triazine
       Thiols, *J. Soc. Rubber Ind. Jpn.*, *57*, 376 (1984).

43.   Y. Nakamura, M. Saito, K. Mori, and Y. Asabe, Coupling Ef-
       fects of Trithiocyanuric Acid for Vulcanizing Adhesion of Sty-
       rene-Butadiene Rubber on Copper Plate, *Kobunshi Ronbunshu*,
       37, 389 (1980).

44.   Y. Nakamura, K. Mori, and K. Tamura, Coupling Effect of
       6-Dibutylamino-1,3,5-triazine-2,4-dithiol for Adhesion of Poly-
       ethylene on Copper Plate, *J. Adhes. Soc. Jpn.*, *17*, 308 (1981).

45.   K. Mori and Y. Nakamura, Adhesion of Soft PVC Sheets to Ny-
       lon Fabrics, *Kobunshi Ronbunshu*, *35*, 375 (1978).

46.   Y. Tsukahara, K. Kohno, H. Inoue, and Y. Yamashita, Surface
       Modification of Polymer Solids by Graft Copolymers, *Chem. Soc.
       Jpn.*, *6*, 1070 (1985).

7

Adhesion of Resin to Metal

7.1 INTRODUCTION

The surfaces of solids generally adsorb CO_2, O_2, N_2, and H_2O from the atmosphere and are frequently covered with molecular films of fats and oils. When such a metal is mechanically polished or cleaned with an organic solvent for degreasing purposes, wetting of the adhesive, or adhesion, can be improved. Although the contact angle between the metal surface and water just after cleaning is approximately 70 to 90°, showing lipophilic characteristics, an oxidized film is produced on it in an atmosphere that is hydrophilic in nature [1]. For instance, an oxide layer such as Fe_2O_3 and Fe_3O_4 is produced on the surface of iron. Such a layer spontaneously developes rust, which prevents adhesion because it is brittle and contaminated. The rusty surface of metal is therefore mechanically polished or pickled to form continuous, active, tough oxide film by further surface treatment. Recently, quantitative measurement of the adhesion properties of polymer films to metals has also progressed [2,3].

7.2 ADHESION OF PLASTIC TO METAL

7.2.1 Metal Surface and Adhesion

Levine et al. [4a,4b] investigated the contact angle, roughness of surface, and adhesive strength under shear by binding epoxy resin to various surface-treated steels. The results, shown in Table 7.1, reveal that the contact angle is always reduced, and consequently the adhesive strength is increased, by treating the steel with a degreasing solvent. By treating the steel with chromic acid of pH 0.1 or less or 50% HNO_3 for 10 s, the contact angle is reduced to 30 to 38° and the surface roughness becomes 10 to 20 μm. The adhesive strength, however, ranges from 104 to 130% of the base strength of steel treated with trichloroethylene.

Table 7.1   Relationship Between Surface Treatment, Contact Angle and Tensile Shear in an Epoxy Adhesive System[a]

Surface treatment	Contact angle (deg)	Tensile Shear (psi)	Surface roughness (μin.)
Effect of Solvent Treatment			
As received[b]	77	500	10-15
Standard solvent cleaned (Trichloraethylene)	42	1940	10-15
Additional Solvents			
Toluene	59	1795	10-15
Heptane	51	1800	10-15
Methyl ethyl ketone	47	1805	10-15
Ethyl acetate	43	1945	10-15
Methyl chloroform + treatment in ultrasonic vibrator			
5 min	35	2140	10-15
13 min	34	2210	10-15
20 min	34	2190	10-15
Effect of Mechanical Treatment[c]			
Sisal buffing (30 s)	42	2000	5
Polishing (30 s + Sisal 1 min)	42	1980	5
Polishing (1 min)	41	2100	10
Buffing (1 min)	44	2000	10-15
Grit blasting	36	2425	80-100
Effect of Chemical Treatment[d]			
Chromic acid: pH 0.6-0.8	42	1960	10-20
Chromic acid: pH < 0.1	38	2170	10-20
Hydrochronic acid diluted to 50% with water			
1.0 min	38	2008	10-20
3.5 min	37	2128	10-20
7.5 min	35	2215	10-20

Table 7.1   (Continued)

Surface treatment	Contact angle (deg)	Tensile shear (psi)	Surface roughness (μin.)
Effect of Chemical Treatment[d]			
Nitric acid diluted to 50% with water			
1 s	38	2320	10-20
5 s	35	2500	10-20
10s	35	2545	10-20
Sulfuric acid sodium dichromate etch			
0.5 min	34	2060	25-30
2.0 min	34	2020	40-50
5.0 min	34	2100	60-70
Hydrofloric acid			
1.0 min	29	2170	10-15
2.5 min	29	2240	10-15
Alkaline etch pH 12.6, 180°F, 10 min	36	2320	10-20

[a]Adhesive; Epoxy resin (Epon 828)/methane diamine/Versamide 115 = 100:22.5:20

[b]Adherend; Steel (SAE 1010), 10 ∿ 15 μin surface finished, stored in trichloroethylene for 4-5 h at room temp., hand wiped and rinsed in clean trichloraethylene, dried with a hot air gun.

[c]Standard solvent-cleaned followed by mechanical treatment and again solvent cleaning.

[d]Standard solvent-cleaned followed by chemical treatment rinsing with water and drying with hot-air gun.
Source:Ref. 4a.

Upon dipping aluminum into water, $\beta-Al_2O_3 \cdot 3H_2O$ (bayerite) and $\alpha-Al_2O_3 \cdot H_2O$ (boehmite) are produced at temperatures, respectively, of below 160°F (N71°C) and above 160°F. It is said that a stable amorphous oxide layer, $\gamma-Al_2O_3$, is transformed into extremely adhesive $\beta-Al_2O_3 \cdot 3H_2O$ by treating the surface with a mixture of chromic acid and sulfuric acid or by anodizing in phosphoric acid and chromic acid solutions. The microscopically rough structure of the dense oxide layer formed by these treating methods develops a high mechanical bonding strength with resin, including epoxy resin. In addition, it improves the wetness of the polymer material by removing impurities such as hydrocarbon adsorbed on the metal surface. The OH group on the oxide surface is chemically bonded to the polymer material. It is believed, for example, that an extra epoxy group in the epoxy resin causes the following reaction in the presence of a proper catalyst [5]:

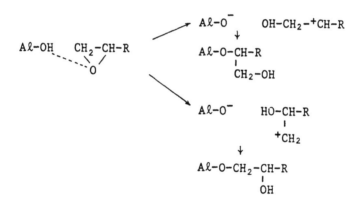

Moisture in the atmosphere acts on the oxide layer to produce flaky aluminum hydroxide ($Al_2O_3 \cdot 2H_2O$), which lessens adhesion to the Al substrate, as shown in Figure 7.1. It has been found that nitrile trismethylene phosphate (NTMP) is effective as an inhibitor of that reaction. NTMP is adsorbed on the surface of the oxide from the aqueous solution and a P—O—Al bond is formed by substituting a hydroxyl group as shown in Figure 7.2 [6]. Mechanical bonding of the oxide layer with polymer material, chemical bonding on the interface, and stability of the oxide layer are required to increase the adhesive strength and durability of the polymer material bound to the metal surface.

The surface of copper is generally covered with an oxide film, CuO, in the atmosphere and the sublayer is the diffusion layer of Cu_2O. Since Cu_2O contributes to adhesion, it is necessary to form an active film of Cu_2O by any proper surface treatment method. Vazirani's surface-treating method [7] is as follows. Copper is etched in an acidic bath of concentrated phosphoric acid/concentrated nitric acid/water (75:10:15) for 30 s, or low-copper alloy is dipped into 20% NaOH aqueous solution at 200°F ($\approx$ 93°C) for 5 min. The low-

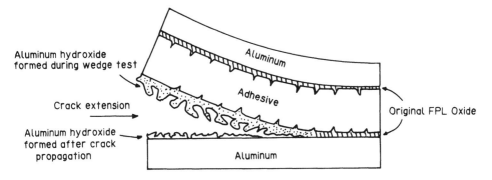

Figure 7.1 Schematic drawing of the mechanism deduced from crack propagation during wedge testing. In a humid environment, the original oxide is converted to a hydroxide that adheres poorly to the aluminum substrate. The crack propagation rate is faster here than in a dry atmosphere, where the crack propagates directly through the adhesive. (Ref. 6.)

copper alloy is treated in an aqueous solution containing 1% each of permanganate and NaOH at 200°F for 5 min accompanied by washing with water between each treatment. The result is copper covered with a continuous, mechanically tough, thin (500 to 1000 Å) oxide film without corrosion. Adhesives that attack copper reduce the durability of adhesion. For instance, a certain adhesive of silicon-RTV

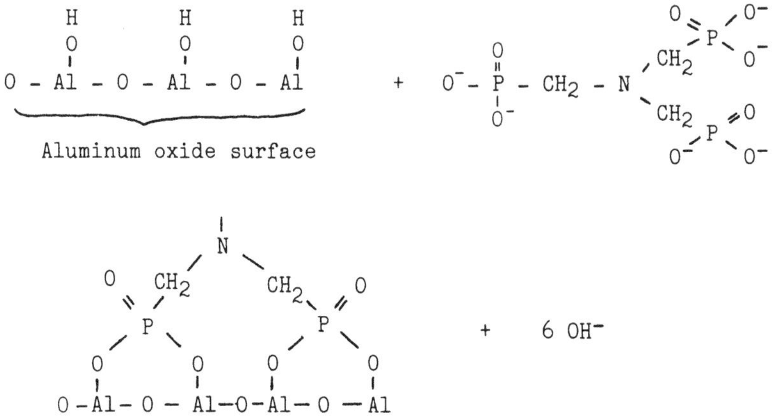

Figure 7.2 Model for the adsorption of NTMP onto aluminum oxide surfaces. The deprotonated NTMP molecule replaces adsorbed hydroxyl ions, resulting in P—O—Al bonding. (Ref. 6.)

silicon-RTV evidences acetic acid at curing, and ethylene-acrylate copolymer produces a salt with copper under certain conditions. To prevent the copper from corroding, after acidic bath etching as described above, electrolytic oxidation is carried out in an electrolyte composed of 3% each sodium chromate, sodium citrate, and sodium carbonate at a current density of 10 A/ft^2 for 1 min using a stainless steel anode and a cathode of the copper being treated. By this method, uniform, corrosion-resistant chromium oxide film as thin as approximately 25 Å containing as little as 1.0 µg Cr/cm^2 as determined by x-ray spectrometry and atomic-absorption spectroscopy is formed. This oxide film is useful for improving adhesive strength and durability.

Copper scarcely adheres to polyethylene in general. According to Bright et al. [8], this is because copper inhibits the oxidation of polymer. The thick, black-matted oxide film formed by oxidizing copper improves adhesion. The oxide can be produced from a solution of sodium chlorite, sodium triphosphate, or sodium hydroxide. Using coulometry, Baker et al. [9a] analyzed the oxide on the surface of copper before and after hot-melt coating with polyethylene (PE) to examine the adhesion to polyethylene of oxidation-treated copper. The results indicate that cupric oxide, CuO, is decreased, and cuprous oxide, Cu_2O, is increased, in the oxide film adhering to polyethylene. It is believed that cupric oxide oxidizes polyethylene to improve adhesiveness as shown in the following formula:

$$2CuO + PE \longrightarrow Cu_2O + PE\{0\} \tag{1}$$

Meanwhile, in work by Evans et al. [9b] it was estimated from the analytical results of CuO and Cu_2O on the surface of copper sheets with and without hot-melt coating of polyethylene carried out at 200°C for 20 min under optimal conditions for copper-polyethylene adhesion that Cu_2O is produced according to the following reaction rather than the interaction of CuO with polymer:

$$CuO + Cu \longrightarrow Cu_2O$$

$$4Cu + O_2 \longrightarrow 2Cu_2O \tag{2}$$

Peel strength is greatly enhanced by a presputtering treatment of the Teflon prior to the deposition of Cu. Without such a treatment, the Cu-Teflon showed a peel strength of less than 1g/mm, and the Cu film can easily be removed by Scotch tape. The peel strength increases by more than 20 times at 10 sec of presputtering, and reaches 50g/mm after 30 sec of sputtering. The peel strength remains nearly constant at longer sputtering times. It is shown that, for a finite chemical bonding, an appreciable contribution to the peel strength is possible from the morphology changes observed [10].

7.2.2   Roughness of Metal Surface and Adhesion

Although the relationship between the roughness of a metal surface and
the adhesion attainable is given in numerous literature references, it
is not necessarily reproducible. This is because adhesive failure is
not reproducible: that is, it does not always take place on the interface, although local cohesive failure frequently takes place there. Mechanical surface treatment to roughen the surface is therefore carried
out to enlarge the specific surface area, to form a cleaned and highly
active surface, and to control rapid brittle fracture in the shear or
fracture test.

When the adhesive penetrates voids in the rough surface, the
penetrating depth can be expressed by the following equation, based
on the theory of capillarity:

$$h = k\gamma_{L\upsilon} \frac{\cos \theta}{\rho R} \qquad (3)$$

where $\gamma_{L\upsilon}$ is the surface tension of the liquid adhesive, θ the contact angle representing wetness of adhesive, ρ the density of the
adhesive, and R the radius of the capillary tube.

If we represent the factor of adhesive area increase due to roughness by γ, that of loss due to void by g, and that of stress concentration by S, the apparent adhesion work, Wa, is expressed by γ/gS.
Assuming that liquid is moved along the distance dx on the rough
surface, the interfacial area between solid and liquid and that between gas and liquid are increased or decreased in proportion to
γ dx. We thus have the equation

$$\gamma(\gamma_{s\upsilon} - \gamma_{sL}) = \gamma_{L\upsilon} \cos \theta' \qquad (4)$$

where γ is the coefficient of roughness, a correction term for surface roughness, and θ' is the contact angle on a rough surface. Assuming that the surface of a solid does not adsorb the vapor of a
liquid, as mentioned in the preceding section, we have

$$\gamma_{s\upsilon} = \gamma_{sL} + \gamma_{L\upsilon} \cos \theta \qquad (5)$$

From equations (4) and (5), the equation expressing the relationship
between the surface roughness and the contact angle is

$$\gamma = \frac{\cos \theta'}{\cos \theta} \qquad (6)$$

When the metal surface is roughened, it becomes easy to wet because the apparent constant angle θ' between the liquid and the metal
surface becomes smaller than the real contact angle, θ. Since the
surface roughness where the most effective adhesion is realized

Table 7.2 Surface Preparation Versus Lap Shear Strength[a]

Group treatment	$\overline{X}$ (psi)	s (psi)	C_V (%)
1 Vapor degrease, grit blast 90-mesh grit, alkaline clean, $Na_2Cr_2O_7-H_2SO_4$, distilled water	3091	105	3.5
2 Vapor degrease, grit blast 90-mesh grit, alkaline clean, $Na_2Cr_2O_7-H_2SO_4$, tap water	2929	215	7.3
3 Vapor degrease, alkaline clean, $Na_2Cr_2O_7-H_2SO_4$, distilled water	2800	307	10.96
4 Vapor degrease, alkaline clean, $Na_2Cr_2O_7-H_2SO_4$, tap water	2826	115	4.1
5 Vapor degrease, alkaline water, chromic$-H_2SO_4$, deionized water	2874	163	5.6
6 Vapor degrease, $Na_2Cr_2O_7-H_2SO_4$, tap water	2756	363	1.3
7 Unsealed anodized	1935	209	10.8
8 Vapor degrease, grit blast 90-mesh grit	1751	138	7.9
9 Vapor degrease, wet and dry sand, 100 + 240-mesh grit N_2 blown	1758	160	9.1
10 Vapor degrease, wet and dry sand, wipe off with sandpaper	1726	60	3.4
11 Solvent wipe, wet and dry sand, wipe off with sandpaper (done rapidly)	1540	68	4
12 Solvent wipe, sand (not wet and dry), 120 grit	1329	135	1.0
13 Solvent wipe, wet and dry sand, 240 grit only	1345	205	15.2
14 Vapor degrease, aluminum wool	1478	—	—
15 Vapor degrease, 15% NaOH	1671	—	—
16 Vapor degrease	837	72	8.5
17 Solvent wipe (benzene)	353	—	—
18 As received	444	232	52.2

[a]Resin employed in Epon 934 Shell Chemical Company; cured 16 h at 75°F plus 1 h at 180°F. Fillet on overlap left intact, adhesive on sides of specimens removed.
Source: Ref. 11.

generally depends on the combination of adherend and adhesive, the adhesive conditions must be examined closely. According to Chessin et al. [11], the adhesive strength under shear in the adhesion system aluminum/Epon 934 is increased by increasing grit size from 90 mesh to 24 mesh in the order chromic acid mixture treatment $\gg$ grit blasting = grit blasting + chromic and mixture treatment > mechanical abrasion + chromic and mixture treatment. The results are shown in Table 7.2.

Rogers reported that the adhesive strength under shear in an aluminum/modified epoxy resin system is increased more by treatment with sharp grains such as quartz powders [12] than by treatment with dull grains such as glass beads. Few reports on the quantitative relationship between surface roughness and adhesive strength have been published.

Malpass et al. [13,14] reported on the relationship between oxide film thickness, peel strength under shear, and the void by measuring the diameter of hexagonal-prism pores (120 Å in H_2SO_4 bath and 330

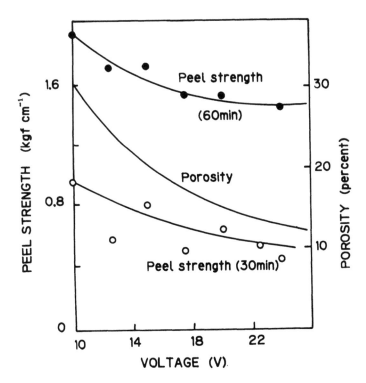

Figure 7.3 Variation in peel strength and porosity with anodizing voltage for films formed in phosphoric acid at approximately constant current density. Anodizing times: ●, 60 min; ○, 30 min. (Ref. 14.)

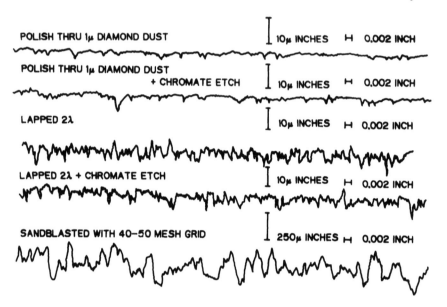

Figure 7.4   Talysurf profilometer traces for 304 stainless steel surfaces. (Ref. 15.)

Å in H_3PO_4 bath) regularly formed on anodized aluminum film. The results are shown in Figure 7.3.

Jenning et al. [15] published diagrams of the surface roughness (measured using a roughness meter) of aluminum and stainless steel abraded chemically or mechanically (Figure 7.4) and of the relationship between adhesive strength under tension and temperature characteristics (Figure 7.5).

7.2.3   Adhesiveness of Metallic Fiber to Matrix in Metallic Fiber-Reinforced Plastic

Few reports on the adhesiveness of metallic fiber to matrix have been published, although there are many reports on the use of metallic fiber to reinforce plastic [16-18]. Surface treatment of metallic fiber-reinforced plastics includes (1) cleaning the metal surface, (2) roughening the metal surface, (3) applying oxide film, (4) applying chemical conversion coatings, and (5) coating with another metal. The effect of surface treatment of metallic fiber on the adhesiveness to epoxy resin of steel wire treated by various methods according to McGarry et al. [16] is shown in Figure 7.6. As the figure shows, treatment by phosphoric acid is extremely effective.

The mechanical properties of laminated sheets are shown in Table 7.3. The table reveals, however, that the effect of coating with

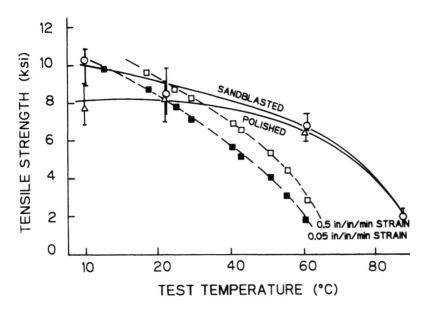

Figure 7.5   Variation of joint strength with temperature for Epon
815-Versamid 140 (60/40) adhesive and 6061 T6 Al adherends.   Dashed
lines indicate bulk polymer data of Ishai.   (Ref. 15.)

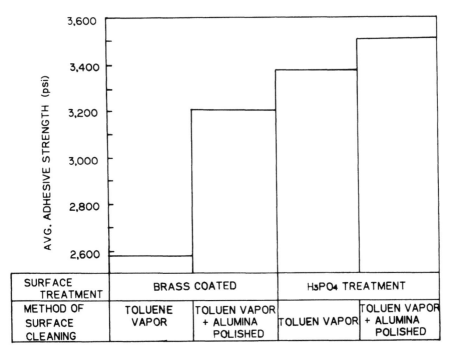

Figure 7.6   Adhesive strength of steel and epoxy resin (Epon 828Z).
(Ref. 16.)

Table 7.3   Mechanical Properties of Unidirectionally Fiber-Reinforced Epoxy Resin

Treated fiber V_f (%)	H_3PO_4-treated steel 70	Brass-coated steel 70	Glass fiber 84 (wt%)
Epoxy resin	Epon 828 (50 phr) + Araldite 6005 (50 phr)	Epon 828 (50 phr) + Araldite 6005 (50 phr)	Epon 828
Tensile modulus (psi) 10^6	19.7	21.2	7.5
Flexural modulus (psi) 10^6	12.8	15.2	7.4
Flexural strength (psi) 10^5	2.46	3.18	2.19

Source: Ref. 16.   The wires used had a tensile strength of 575,000 psi.

brass is greater than that noted above. This maybe because the brass coating, acting as an intermediate layer, is effective in transferring power between the fiber and the resin.

7.3 ADHESION OF ELASTOMER TO METAL

7.3.1 Adhesion of Rubber to Metal

The adhesion of elastomer to metal is carried out by both direct and indirect methods [19]. The direct method is to bond the metal to the rubber surface during vulcanization of the elastomer, while the indirect method is to attach adhesive-applied metal to the elastomer with the adhesive between. The metals used in the direct method include brass (copper/zinc 70:30) and cobalt. Zinc can only be adhered to elastomer to which organic cobalt salt has been added. Steel cord plated with brass adheres firmly to elastomer by the direct method.

In this section we describe the adhesion to metal of natural rubber, the most commonly used elastomer. Using the direct method, brass-plated metal is laid on top of unvulcanized rubber and they are heated in a press. The cross-linking reaction of rubber and the bonding between rubber and brass progress competitively, resulting in adhesion. It is believed that various reactions (Figure 7.7) take place at the same time inside the rubber and on the interface between the rubber and the brass during heating [20]. Since

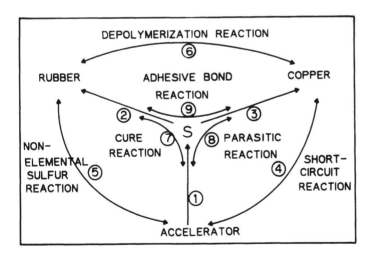

Figure 7.7 Hypothetical scheme of the reactions occurring during the cure of brass in rubber. (Ref. 20.)

the carbon black reinforcing filler, stearic acid vulcanization aid, and zinc oxide compounding additive affect those reactions, and organic cobalt salt and Hisil-resorcin hexamethylene tetramine (HRH) accelerate the adhesive reaction, the factors participating are diverse and complex. Since the metal surface is plated with brass, the composition and structure of the brass layer close to the surface is a concern.

There is an extremely thin (1-nm) film of oxides of copper and zinc underlain by the layer of zinc oxide containing copper on the surface of brass (copper/zinc 70:30), covering the block of copper and zinc [21]. This adhesive interface between the brass and the rubber forms an interfacial layer approximately 100 nm thick, as shown in Figure 7.8 [22]. This layer is double, consisting of a zinc oxide layer containing a small quantity of zinc sulfide on the metal side and a cuprous sulfide layer on the rubber side. Adhesion therefore takes place between the rubber surface and the brass surface covered with cuprous sulfide. Sulfur exists in the form of a cross-link, $-C-S_x-$ $-C-$, and S^{2-} in the interfacial layer. The spectra of bound energy of copper and sulfur shown in Figure 7.9 [22] reveal that cuprous sulfide on the adhesive interface is an nonstoichiometric compound, $Cu_{1.97}S$, because it contains more sulfur than Cu_2S. The spectrum

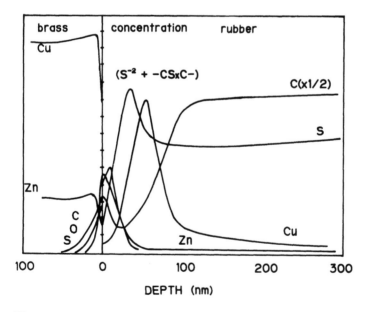

Figure 7.8 In-depth concentration profiles at a rubber-brass interface. Vulcanized 25 min at 150°C. Rubber-contained cobalt-based adhesion promotor. (Ref. 22.)

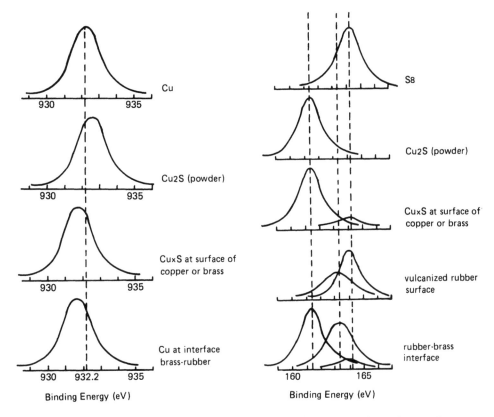

Figure 7.9   Cu $2p_{2/3}$ and S_{2p} photo line in some selected samples.
(Ref. 22.)

of S_{2p} indicates that the C—S bond/free S ratio of 5 is higher than that inside vulcanized rubber. This is because the cross-linking of sulfur is accelerated by catalytic reaction of Cu_xS on the interface, and the concentration of reticulate chain is increased. Since the modulus of the adhesive interface inside the rubber becomes higher with decreasing distance to the metal surface, the stress dispersion improves adhesion.

As a result, the reaction on the interface between brass and rubber during the vulcanization is believed to progress as follows. The copper in the brass is separated by anodic reaction into an electron and a Cu^{2+} ion, which penetrates the zinc oxide layer and diffuses to the surface to produce Cu_xS containing ZnS. Since the reaction ratio of sulfuration is fast, the diffusion rate in zinc oxide is the rate-determining step. Sulfur molecules are adsorbed on the film of Cu_xS, which is dissociated, and part of Cu_xS is desorbed to cross-link with rubber molecules. In the midst of the adsorption process, the sulfur molecules are reduced by cathodic reaction to S^{2-} ions, which are reacted with Cu^+ ions to grow Cu_xS film.

The adhesion is therefore the physical adsorption of cross-linked reticulate chain sulfur in the vulcanized rubber on the Cu_xS surface. It is believed that the adhesiveness, although relatively low, becomes greater than the tensile strength of rubber with increasing cross-link density. Since the Cu_xS film becomes porous during formation, the anchoring effect of the rubber molecules entering the pores helps to increase the adhesiveness [23].

Failure of a rubber-brass adherend is classified into five types: (1) cohesive failure of the rubber, (2) adhesive failure at the sulfide-rubber interface, (3) cohesive failure of the sulfide layer (4) adhesive failure at the sulfide-ZnO interface, and (5) adhesive failure at the ZnO-metal interface. Since these local failures depend on the methods of treating the metal surface and rubber and the compounding additives, how local failure causes adhesion to deteriorate in an aging or corrosive environment must be considered.

7.3.1 Factors Affecting Adhesion of Rubber

Surface Composition of Brass

Since the diffusion rate of copper toward the surface is controlled by the zinc oxide layer that covers the brass surface, the layer of zinc oxide required to provide good adhesion must be of the proper thickness. In this case the brass plating contains 60 to 70% copper. The degree of adhesion is lessened by heating in air because the zinc oxide in the surface layer is increased and the copper is greatly reduced [24].

Rubber Compounding Additives

Effect of vulcanization aid. Since the sulfur cross-linking in rubber and sulfuration of copper take place simultaneously by direct adhesion during the relative reaction rate determines the level of adhesiveness. Since the adhesiveness reaches a maximum before the completion of sulfur cross-linking in the vulcanization of brass-plated steel cord, as shown in Figure 7.10, sulfene amide vulcanization accelerators are used only for reactive brass that contains a large quantity of copper. A super accelerator such as tetramethylthiuram disulfide lowers adhesiveness considerably because cross-linking has been completed and sulfur has been consumed before the sulfuration of copper begins. Since the high sulfur content improves adhesion by increasing the modulus of rubber and the sulfur content used for

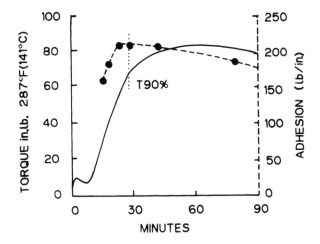

Figure 7.10   Relationship between state of cure and adhesion.   (Ref.
25.)

the sulfuration of copper is also increased, adhesiveness is increased
by retarding the competitive reaction to cross-linking [25,26].

Together with the vulcanization accelerator, zinc oxide and stearic acid accelerate the cross-linking of sulfur. When the content of
zinc oxide in rubber is low, zinc oxide is removed from the brass
during vulcanization. The adhesiveness is therefore reduced because
the decreased zinc oxide layer in the brass cannot retard the sulfuration of copper. As a result, the degree of dispersion and morphology
of the zinc oxide, one of the compounding additives of rubber, also
affects the level of adhesion [27]. Since stearic acid reacts with
zinc to accelerate the dezincification of the brass surface, the adhesiveness is reduced [28].

Adhesive accelerator. Corrosion resistance and the adhesion of
natural rubber to brass plating are generally improved by adding an
adhesive accelerator such as organic cobalt salt and HRH though natural rubber is directly adhered. The organic cobalt salt includes cobalt naphthenate, cobalt octylate, and cobalt stearate. The degree
of adhesiveness is lowered remarkably by adding cobalt octylate
during high-temperature vulcanization with increased vulcanizing
time, as shown in Figure 7.11. It is hardly noticeable from the variation in tensile strength with vulcanizing time that this is caused
by the deterioration of rubber through the addition of cobalt octylate [29].

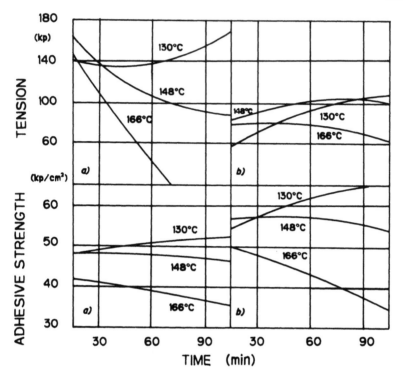

Figure 7.11   Effect of vulcanizing time and temperature on adhesive
strength and tension: (a) blank, (b) cobalt octylate added. (Ref. 29.)

Based on the theory that adhesion is caused by the chemical bonding of cuprous sulfide with sulfur in cross-linked rubber, adhesiveness is reduced through the following reactions in vulcanization [30]:

1.   The vulcanization accelerator, VA, produced during vulcani-
     zation by the addition of zinc oxide and a fatty acid produces
     a complex, $VA-S_x$, that reacts with sulfur, which in turn
     is reacted with a cobalt salt to produce a cobalt complex.
2.   The cobalt complex promotes the production of cuprous sul-
     fide.
3.   The cuprous sulfide causes the bonding of rubber, R, with
     metal.
4.   The cuprous sulfide is converted to cupric sulfide in a much
     higher concentration of sulfur, or by adding too much vulcan-
     ization accelerator or cobalt salt, and the degree of adhesive-
     ness is reduced.

5. The cobalt salt accelerates the sulfuration of copper in the brass as well as that of copper bonding with rubber by heating for a longer period, thus breaking the bonding [31].

$$VA-S_x + CoX_2 \longrightarrow Co-S_z-X_2 + Va-S_{x-z} \tag{7}$$

$$VA-S_x + 2Cu \longrightarrow Cu_2S + VA-S_{x-1} \tag{8}$$

$$Cu_2S + R \xrightarrow{\quad Co-S_x-X_2 \quad} R-S-Cu \tag{9}$$

$$Cu_2S + VA-S_x \longrightarrow 2CuS + VA-S_{x-1} \tag{10}$$

$$2R-S-Cu \xrightarrow{\quad Co-S_x-X_2 \quad} 2CuS + R-S_y-R \tag{11}$$

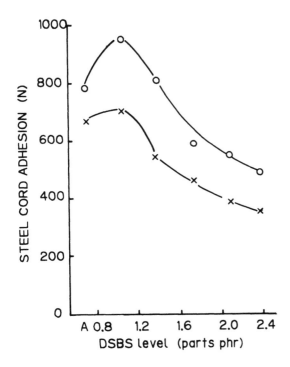

Figure 7.12   Effect of N,N-dicyclohexylbenzothiazole-2 sulfenamide (DCBS) as accelerator on adhesion: $\circ$, unaged; $\times$, steam aged. (Ref. 32.)

Aging

It is known that the adhesiveness of rubber to a brass surface is reduced by the nature of the atmosphere during aging [31,32], as
shown in Figure 7.12. Water corrodes zinc in a high-temperature,
high-humidity atmosphere, the time leaving the brass surface in the
form of zinc oxide and/or zinc hydroxide. As a result, a large quantity of cupric sulfide is produced on a brass surface that has a higher copper content. Because the grating space of cupric sulfide is
larger than that of the substrate brass, failure can occur on the
interface between cupric sulfide and brass by loss of adhesive power
due to the difference in grating spaces. Although cobalt salt prevents brass from dezincifying, adhesion is lowered with aging of
the rubber because in high-temperature aging, rubber is oxidized
by the oxidative catalytic reaction of cobalt salt.

It has been reported that a polyborosilicate adhesive aid improves
adhesion after steam aging and serves to retain adhesion in a salt
corrosion atmosphere because it improves the thermal stability of the
sulfide [33]. The effects of the addition of cobalt salt on the adhesion of rubber to zinc and its adhesion mechanism have been
studied [34,35].

7.4 MODIFICATIONS OF METAL SURFACE BY TRIAZINE THIOLS

Triazine thiols are tautomers between thionyl and thiol. It is said that
thionyl predominates in the solid state, while thiol predominates in the
liquid state and almost all of them participates in the reaction. The
dissociation constant, pK_{a1}, of the thiol group is approximately 4 to
6.5 and the pK_{a2} value is approximately 12. Since thiols are acidic
materials such as hydrogen sulfide, they react readily with metals and
their oxides to produce stable metal oxides. Triazine thiols are generally soluble in organic solvents, while their sodium salts are water
soluble. The surface treatment of metal is carried out in such a way
that the metal is dipped in the organic solvent solution of triazine thiol
or in the water solution of its sodium salt at a concentration of 10^{-4} to
10^{-2} mol/liter at a temperature of 20 to 150°C for 30 s to 60 min. Variously functioning triazine thiol film is produced on the metal surface
by this method.

The characteristics of triazine thiols are shown in Table 7.4 [36].
The table reveals that the amount of film produced varies widely according to the number of carbon atoms in the triazine thiol substituent
group (R). The formation of organic thin films on various metal surfaces such as copper, nickel, zinc, lead, brass, cupronickel and stainless steel by using 6-substituted 1,3,5-triazine-2,4-dithiol monosodium
salts (RTDN) occurs by the diffusion and the build-up mechanisms which
vary with the carbon number of 6-substituents in RTDN. By selecting

Table 7.4   Typical Triazine Dithiols (RTD) and Their Abbreviatied Names

Abbreviated name	M	—R
TT(N1)	H(Na)	—SH
DM(DMN)	H(Na)	$-N(CH_3)_2$
AF(AN)	H(Na)	$-NHC_6H_5$
DB(DBN)	H(Na)	$-(C_4H_9)_2$
DO(DON)	H(Na)	$-N(C_8H_{17})_2$
DL(DLO)	H(Na)	$-(C_{12}H_{24})_2$
DA(DAN)	H(Na)	$-N(CH_2CH=CH_2)_2$
OL(OLN)	H(Na)	$-NHC_8H_{16}CH=CHC_8H_{17}$

Source: Ref. 36.

various functional groups as 6-substituents in RTDN, it will be possible to endow various functions onto metal surfaces [37]. Maybe this is because the above production mechanism of film from RTDN with a substituent group that has 10 or more carbon atoms is different from one that has fewer than 10 carbon atoms. The results are shown in Figure 7.13. The effects of metal type on film production are shown in Table 7.5 [37]. The table reveals that the production rate of film depends on triazine thiol type. Triazine thiols are characterized by reaction with various organic materials. Figure 7.14 shows the cross-linking reaction curves of unsaturated elastomer by triazine thiol type [38]. The figure reveals that most elastomers show practical cross-linking reactions. Triazine thiols are used as curing agents or modifiers, reacting with epoxy resin, unsaturated polyester resin, and urethane resin; they also act as cross-linking agents and modifers, reacting with a polyolefin such as polyethylene in the presence of peroxide.

Figure 7.15 shows the initial peel strength and the peel strength after heat treating of ethylene-propylene-diene terpolymer (EPDM) adhering to brass plates treated with polyethylene glycol solutions of three triazine thiols with different physical and chemical properties [39]. These peel strengths clearly vary according to

Table 7.5   Effect of Metals on the Formation of Thin Films with DMN and DLN onto Metal Surfaces[a]

Metal	k for DMN $\left(\dfrac{\mu mol}{min^{0.5} \cdot dm^2}\right)$	k for DLN $\left(\dfrac{\mu mol}{min \cdot dm^2}\right)$
Copper	1.90	0.27
Nickel	b	0.11
Zinc	b	c
Lead	5.87	c
Brass	1.84	c
Cupronickel	0.38	0.30
Stainless	b	0.07
Polyethylene	0	0.07

[a]Conditions: temperature, 80°C; time, 30 min; concentration, 10^{-3} mol/liter.

[b]Weight did not gain.

[c]Weight lost.

Sources: Ref. 37.

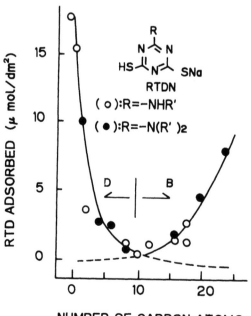

NUMBER OF CARBON ATOMS

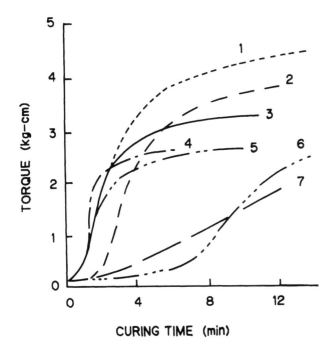

Figure 7.14 Curing of elastomer with TT (2 phr) and CBS (6 phr) in the presence of ZnO (5 phr) and stearic acid (2 phr) at 160°C. 1, Br: 2, EPDM; 3, SBR; 4, NBR; 5, IR; 6, Parel; 7, AR. (Ref. 38.)

Figure 7.13 Effect of the number of carbon atoms of the substitute groups (R) in RTDN on the amount of RTD adsorbed to copper plates at 80°C for 30 min, concentration: 10^{-3} mol/l, D: diffusion mechanism, B: build-up mechanism. (Ref. 37.)

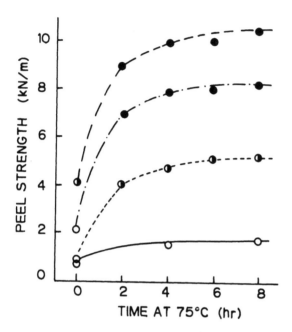

Figure 7.15 Effect of triazine thiols on peel strength in the vulcaniz-ing adhesion of EPDM to brass plates treated with the polyethylene glycol solution (100 ml) of triazine thiols (0.3 g) at 130°C for 15 min. —, blank; ⋯⋯, DB; —·—, AN; — —, TT. ○, interface failure; ●, cohesive failure; ◑, both interface failure and cohesive failure. (Ref. 39.)

triazine thiol type, increasing in the order triazine thiol acrylonitril > DB > blank. The order is the same as those for film thickness and the pK_a of the SH group. Triazine thiol treatment of metals improves the heat resistance, water resistance, and oil resistance of the com-posite materials as well as the adhesiveness.

REFERENCES

1a. F. E. Bartell and L. S. Bartell, Quantitative Correlation of Interfacial Free Surface Energies, *J. Am. Chem. Soc.,* 56, 2205 (1934).

1b. W. D. Harkins and G. E. Boyd, The Binding Energy Between a Crystalline Solid and a Liquid, *J. Am. Chem. Soc.,* 64, 1195 (1942).

2. M. Allen and S. Senturia, Polyimide-Metal Adhesion Measurement Using Microfabricated Structures, 997, SPI, *Proc. 46th Annual Tech. Conf. & Exib.*, Atlanta, (1988).

3. M. Frkumura, Adhesive Bonding of Metals, *Metal Surf. Tech.* *30*, 171 (1979).

4a. M. Levine, G. Ilkka, and P. Weiss, Wettability of Surface Treated Metals and the Effect on Lap. Shear Adhesion, *Adh. Age*, 7, No. 6, 24 (1964).

4b. D. D. Eley, *Adhesion Fundamentals and Practice*, University of Nottingham (McLaren & Sons. Ltd., London), 1969.

5. E. H. Andrews and N. K. King, Adhesion of Epoxy Resins to Metals, *J. Mater. Sci.*, *11*, 2004 (1976).

6. J. D. Venables, Adhesion and Durability of Metal-Polymer Bonds, *J. Mater. Sci.*, *19*, 2431 (1984).

7. H. N. Vazirani, Surface Preparation of Copper and Its Alloys for Adhesion Bonding and Organic Coatings, *J. Adhes.*, *1*, 208 (1969).

8. K. Bright and B. W. Malpass, The Adhesion of Polyethylene to High Energy Substrates, *Eur. Polym. J.*, *4*, 431 (1968).

9a. R. G. Baker and A. T. Spencer, Structural Bonding of Polyolefins, *Ind. Eng. Chem.*, *52*, 1015 (1960).

9b. J. R. G. Evans and D. E. Packham, Adhesion of Polyethylene to Copper, *J. Adhes.*, *9*, 267 (1978).

10a. C. A. Chang, J. E. E. Baglin, A. G. Schrott, and K. C. Lin, Enhanced Cu-Teflon Adhesion by Presputtering Prior to the Cu Deposition, *Appl. Phys. Lett.*, *51*, 103 (1987).

10b. C. A. Chang, Enhanced Cu-Teflon Adhesion by Presputtering Treatment; Effect of Surface Morphology Changes, *Appl. Phys. Lett.*, *51*, 1236 (1987).

11. N. Chessin and V. Curran, Preparation of Aluminum Surface for Bonding, *Appl. Polym. Symp.*, *3*, 319 (1966).

12. N. L. Rogers, Surface Preparation of Metals for Adhesive Bonding, *Appl. Polym. Symp.*, *3*, 327 (1966).

13. B. W. Malpass, D. E. Packham, and K. Bright, A Study of the adhesion of Polyethylene to Porous Alumina Films Using the Scanning Electron Microscope, *J. Appl. Polym. Sci.*, *18*, 3249 (1974).

14. D. E. Packham, K. Bright, and B. W. Malpass, Mechanical Factors in the Adhesion of Polyethylene to Aluminum, *J. Appl. Polym. Sci.*, *18*, 3237 (1974).

15. C. W. Jenning, Surface Roughness and Bond Strength of Ad-
     hesion, *J. Adhes.*, *4*, 25 (1972).

16. F. J. McGarry and D. W. Marshall, Research on Wire-Wound
     Composite Materials, Symposium on Standards for Filamentwound
     Reinforced Plastics, Special Technical Publication *ASTM STP-327*,
     (American Society for Testing and Materials, Philadelphia, 1962.)

17. F. F. Jaray and G. Tolly, A New Reinforcement Material Based
     on Hard Drawn Steel Wire, *21st Annu. Tech. Conf. SPI*, *8-G*
     (1966).

18a. E. Moncunill de Ferran and B. Harris, Compression Strength
     of Polyester Resin Reinforced with Steel Wires, *J. Compos.
     Mater.*, *4*, 62 (1970).

18b. J. W. Sawyer and R. E. Stuart, Ionomer Laminates New Plastic/
     Metal Composites, *Mod. Plast.*, *44*, 125 (1967).

19. M. Ashida, Adhesion of Elastomer to Metal, *6th Symp. Inorg.
     Polym.*, Tokyo, 25 (1986).

20. A. Maeseele and E. Debruyne, Problems Concerning Adhesion
     of Steel Wire and Steelcord in Rubber, *Rubber Chem. Technol.*,
     42, 613 (1969).

21. W. J. Van Ooij, Surface Composition of Commercial Cold-Worked
     Brass, *Surf. Technol.*, *6*, 1 (1977).

22. W. J. Van Ooij, Mechanism of Rubber-to-Brass Adhesion, *Rub-
     ber Chem. Technol.*, *51*, 52 (1978).

23. W. J. Van Ooij, W. E. Weening, and P. F. Murray, Rubber
     Adhesion of Brass-Plated Steel Tire Cords, *Rubber Chem.
     Technol.*, *54*, 227 (1981).

24. G. Haemers and J. Mollet, The Role of the Brass Surface Com-
     position with Regard to Steel Cord Rubber Adhesion, *J. Elasto-
     mers Plast. 10*, 241 (1978).

25. R. C. Ayerst and E. R. Rodger, Steel Cord Skim Compounds;
     The Achievement and Maintenance of Maximum Adhesion, *Rubber
     Chem. Technol.*, *45*, 1497 (1972).

26. K. D. Albrecht, Influence of Curing Agents on Rubber-to-
     Textile and Rubber-to-Steel Cord Adhesion, *Rubber Chem.
     Technol.*, *46*, 981 (1973).

27. G. T. Carpenter, The Effect of Zinc Oxide Particle Size and
     Shape on Adhesion of Rubber to Brass-Coated Steel Radial Tire
     Cord, *Rubber Chem. Technol.*, *51*, 788 (1978).

28. G. Rutz, Uber die Bindung von Nitrilkautschukmischungen an
     Messing Ms 58 ohne Haftmittelzwischenschichten, *Plaste Kautsch.*,
     17, 909 (1970).

29. G. Rutz, Zur zwischenshichtfrein Gummi-Metall-Bindung III, *Plaste Kautsch.*, *23*, 742 (1976).

30. G. Rutz, Zur zwischenschichtfrein Gummi-Metall-Bindung IV, *Plaste Kautsch.*, *24705* (1977).

31. Y. Ishikawa, Effects of Compound Formulation on the Adhesion of Rubber to Brass-Plated Steel Cord, *Rubber Chem. Technol.*, *57*, 855 (1984).

32. L. R. Barker and G. M. Bristow, A. Curing System for Rubber Bonded to Brass-Plated Tire Core, *Rubber Chem. Technol.*, *54*, 797 (1981).

33. W. J. Van Ooij, *Int. Rubber Conf. 84*, Moscow, 97 (1984).

34. M. Ashida, M. Nakatani, and Y. Takemoto, Effect of Cobalt Naphthenate on Adhesion of NR to Zinc Plate, *Rep. Res.*, *Asahi Glass, Found. Ind. Technol. Jpn.*, *41*, 97 (1982).

35. M. Ashida, M. Nakatani, Y. Takemoto, and M. Goto, Studies on Bonding Rubber to Metallic Zinc, *J. Soc. Rubber, Jpn.*, *57*, 544 (1984).

36. K. Mori, Functionality of Metal Surface and Application to Composite, *6th Symp. Inorg. Polym.*, Tokyo, 19 (1986).

37. K. Mori, M. Saito, and Y. Nakamura, *J. Chem. Soc., Jpn*, No, 4, 725 (1987).

38. K. Mori and Y. Nakamura, Curing Reaction of Elastomers with Triazine Thiols and Sulfen Amides, *J. Appl. Polym. Sci./*, *30*, 1049 (1985).

39. K. Mori, Y. Nakamura, M. Shida, and I. Nishiwaki, Vulcanizing Adhesion Mechanism of EPDM to Brass Plates Treated with Triazine Thiols, *J. Soc. Rubber Jpn.*, *57*, 376 (1984).

8

Bonding of Ceramic to Metal

8.1 INTRODUCTION

Despite having excellent properties such as high heat and electric
resistance and high hardness, ceramics have weak points with regard
to mechanical properties and thermal shock. To avoid these, composite materials made of metals and ceramics have been researched
and developed. Such composite materials, which have broad applications in such fields as aggressive environments, high-temperature
engines, marine resource development, space engineering, electron-
ics, and nuclear industries, will be used even more widely in the
future.

Metal-ceramic bonded products are also required to have a strong
internal bond. Moreover, bonding strength influences such factors
as the bonding mechanism associated with phase formation between
metals and ceramics and the boundary structure, the effect of stress
produced by differences in the thermal expansion coefficient, and
the breaking mechanism related to stress. It may be true that existing bonding systems could already be regarded as applicable in
practice, but even for these systems, information on the factors
noted above has not been established.

8.2 METHODS AND BASIS OF BONDING PROCESS

General methods of bonding ceramics with metals include, separately
or in combination, such operations as heating, pressing, and applying voltage. Depending on which phase—gas, liquid, or solid—
accompanies the bonding and which phase predominantly participates
with the bonding, bonding processes may be classified as a

solid-gas phase system, a solid-liquid phase system, or a solid-solid phase system (see Table 8.1) [1].

One of the most troublesome present problems in the bonding of ceramics to metals relates to the residual stress due to thermal expansion that remains the bonded products. Generally speaking, the thermal expansion coefficients of ceramics are lower than those of metals. Typical thermal expansion coefficients of metals and ceramics are shown in Figure 8.1. This difference in thermal expansion produces a high stress within bonded products that have cooled down from a bonding temperature. Tensile stress on the ceramic side often causes the ceramic to break.

A typical means of buffering such internal stress consists of using a soft metal as an intermediate layer. Materials employed include aluminum [2], foamed metal of nickel-based alloy [3], and nickel wire bundle [4]. Putting such a material between the ceramic and the metal as an intermediate layer buffers the strains caused by differences in thermal expansion of the two materials by means of elastic and plastic deformation of the intermediate layer. In addition, there have been many variations on such a bonding scheme, including applying an intermediate layer made of a metal or complex material that has a thermal expansion coefficient ranking between those of ceramic and metal [5]; an intermediate layer with an expansion coefficient close to that of ceramic [6]; or an intermediate layer of a multilayer laminated material.

An additional problem relates to wetting, interface reaction, and material transition on the interface. The reactivity between metals and ceramics depends primarily on temperature. Leaving atmospheric effects aside, an estimate of the reactivity can be obtained by studying the standard free energy of formation of metal by-products of the same origin as elements used in making ceramics, such as oxides, carbides, nitrides, and borates, as well as intermetallic compounds. There are however, some exceptions: Ceramics with a nonstoichiometric rock salt structure, such as TiC and TiN, are energetically stable, but they form mutual solid solutions with iron alloys, Ni, or Mo to form intermetallic compounds and carbides and nitrides on the metal side [7,8]. Some mutual reaction of both materials is necessary to achieve bonding, but too strong a reaction is undesirable. Particularly in the case of a reaction layer, the layer usually requires an optimal thickness value.

In addition, after the reaction, it sometimes happens that for mutual diffusion velocities of very different elements, the Kirkendall effect causes, in addition to solid-phase bonding of the metal, gas bubble formation on the diffusion source side, or a disturbed structure. It is also possible that with oxide or nitride ceramics, gases such as O_2 or N_2, respectively, are generated, depending on the reaction, and as a result, bubbles are formed on the interface. Such bubbles constitute defects in the bonded products.

Table 8.1   Methods for Bonding of Ceramics to Metals/Ceramics

Solid-vapor system	Evaporating method	$\begin{cases} \text{Ceramics-metal} \\ \text{Glass-metal} \end{cases}$
	Ion plating method	$\begin{cases} \text{Ceramic-metal} \\ \text{Glass-metal} \end{cases}$
	Sputtering method	$\begin{cases} \text{Ceramic-metal} \\ \text{Glass-metal} \end{cases}$
	CVD method	$\begin{cases} \text{Ceramic-metal} \\ \text{Glass-metal} \end{cases}$
Solid-liquid system	Method utilizing organic adhesives (1) Epoxide (2) Vinyl acetate etc.	$\begin{cases} \text{Ceramic-ceramic} \\ \text{Glass-glass} \\ \text{Ceramic-glass} \\ \text{Ceramic-metal} \\ \text{Glass-metal} \end{cases}$
	Method utilizing inorganic adhesives (1) Alali silicate (2) Phosphate etc.	$\begin{cases} \text{Ceramic-ceramic} \\ \text{Glass-glass} \\ \text{Ceramic-glass} \\ \text{Ceramic-metal} \\ \text{Glass-metal} \end{cases}$
	Method utilizing oxide solder (1) Noncrystalline solder (2) Crystalline solder	$\begin{cases} \text{Ceramic-ceramic} \\ \text{Glass-glass} \\ \text{Ceramic-glass} \\ \text{Ceramic-metal} \\ \text{Glass-metal} \end{cases}$
	Method utilizing metal solder (1) In and In alloy (2) Al (3) Pb—Sn—Zn—Sb alloy (4) Ti—Ni, Ti—Cu, Zr—Ni, etc.; active alloying method (5) TiH_2—Ni, TiH_2—Cu, ZrH_2—Ni, etc.; hydride method etc.	$\begin{cases} \text{Glass-metal} \\ \text{Ceramic-ceramic} \\ \text{Glass-glass} \\ \text{Ceramic-glass} \\ \text{Ceramic-metal} \end{cases}$
	Sintered metal powder process (1) Mo—Mn process (2) Mo process etc.	Ceramic-metal
	Method utilizing copper sulfide and silver carbonate	Ceramic-metal
	Method utilizing copper oxide	Ceramic-metal
	Direct contact/heating bonding method (1) Heating by furnace or burner (2) Flame spraying etc.	$\begin{cases} \text{Ceramic-ceramic} \\ \text{Glass-glass} \\ \text{Ceramic-glass} \\ \text{Ceramic-metal} \\ \text{Glass-metal} \end{cases}$

Table 8.1   (Continued)

Solid-liquid system	Field-assisted method	$\begin{cases} \text{Glass-metal} \\ \text{Ceramic-metal} \end{cases}$
Solid-solid system	Method for bonding green bodies by direct contact and heating Direct contact/heating bonding method etc.	$\begin{cases} \text{Ceramic-ceramic} \\ \text{Ceramic-metal} \end{cases}$

Source: Ref. 1.

In judging the bond associated with the bonded interface, both crystal coordination between the two materials and bonding styles on the atomic level are considered. The eventual coordination between the materials provides a strong bond. When a reaction layer is formed, it is desirable that it might be in coordination with both

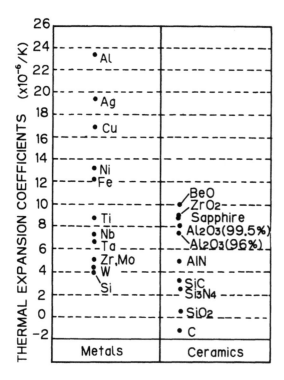

Figure 8.1   Thermal expansion coefficients of metal and ceramic. (Ref. 24.)

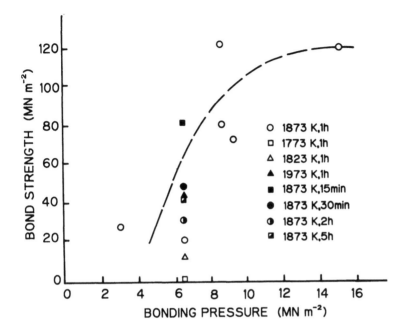

Figure 8.2   Variation of bond strength of Al_2O_3—Nb bond with bonding pressure for specimens heat-treated for 3.6×10^3 s at 1873 K, together with treatment temperature and time dependence of bond strength for the specimen subjected to a stress of 6.4 MN/m^2. (Ref. 9.)

materials. An example of good combined coordination is the bonding of Al_2O_3/Nb, as shown in Figure 8.2, which is well known to bond strongly at temperatures over 1500°C [9,10]. This is due to the fact that NbO_x is formed as a reaction layer, which demonstrates good coordination with Al_2O_3 as well as Nb. A similar observation has been reported for HfO_2/Nb and stabilized ZrO_2/Nb systems [11, 12].

With regard to bonding style on the atomic level, on the ceramic side, ionic bonds, covalent bonds, van der Waals bonds, and so on, are intermingled; on the metal side, metallic bonds are typical. However, on the ceramic side are semiconductor-like bonds or metallic electron state bonds such as SiC, TiC, and TiN. The type of bonding force that functions between the two materials has an important influence on the bonding strength, as well as being scientifically interesting. Typical examples of bonding are described next.

8.3   EXAMPLES OF BONDING PROCESS

8.3.1   Bonding by the High-Melting-Point Metal Process

In a method known as the Telefunken process, Mo, Mo—Mn, or other additives considered effective for metallizing are finely powdered and mixed with an organic binder to form a paint, applied to the ceramic surface, and metallized between 1300 and 1700°C in wetted hydrogen gas or wetted forming gas (H_2/N_2). An important point here is the source of oxygen that contributes to the interfacial reaction. It is reported that owing to the decomposition of base plate ceramics, even thermodynamic considerations would not lead to an oxygen supply. An attempt to accelerate interfacial reaction by the addition of MoO_3 proved to be fruitless, bonding with sapphire (Al_2O_3 monocrystal) proved to require a temperature above 1750°C, and coloring, cracking, and sintering of the ceramic surface were noticeable only above 1900°C [13].

It is believed that the role of added Mn involves MnO formed according to the reaction

$$Mn + H_2O \longrightarrow MnO + H_2$$

reacting with Al_2O_3 in the base plate or SiO_2 included therein as an impurity, forming aluminate or silicate and combining with Mo grains on the ceramic surface [14]. In other words, the Mo—Mn process is said to be useful when including acidic oxides such as those in the base plate or SiO_2 included therein, the MoO_3 addition process being effective when including basic oxides such as CaO and MgO [15].

As mentioned above, when using an Mo—Mn system, the formation of MnO followed by reaction thereof with the base plate or included SiO_2 is expected, although control of the atmosphere is difficult. For this purpose, $MnO-SiO_2-Al_2O_3$ [16–18], $MoO_3-MnO_2-TiO_2-SiO_2$ [19], and $SiO_2-MnO-Al_2O_3-Cr_2O_3-Fe_2O_3-Cu_2O$ [20] systems, for example, were used; other systems, including $CaO-Al_2O_3$ [18,21], $CaO-Al_2O_3-SiO_2$ [18], $MgO-Al_2O_3-SiO_2$ [18,22], $MgO-CaO-SiO_2$ [18], and $MgO-Al_2O_3-SiO_2-Al_2O_3$ [23], have also been employed. Moreover, systems such as $Na_2O-CaO-SiO_2$ and kaolin ($SiO_2-Al_2O_3-H_2O$) [23] have been added to improve the sinterability of Al_2O_3 and MgO.

8.3.2   Bonding by the Active Metal Process

In the active metal process, very active metals such as Ti and Zr, together with Ni, Cu, and Ag, which form alloys that present a relatively low melting point with the active metals, are inserted between ceramics and metals that are to be bonded so as to produce a eutectic

Table 8.2 Composition and
Melting Points of Various Solders

Composition (wt %)	Melting point (°C)
Cu—28 Ti	880
Ni—71.5 Ti	955
Fe—32 Ti	1085
Co—18.6 Ti	1135
Cu—47 Zr	885
Ni—83 Zr	960
Fe—84 Zr	934
Mo—69 Zr	1520
Ni—Ti—Zr	900

Source: Ref. 24.

structure, and are bonded in vacuum or inert gases in a single heat treatment. As shown in Table 8.2, there are various solder compositions, with a wide range of bonding temperatures [24]. For example, the contact angle of an Ti—Ni system with ceramics is saturated within 5 to 10 min, as shown in Figure 8.3. Moreover, analytical results through EPMA of the diffusion aspect of respective elements in the bonded zone of sapphire (α—Al_2O_3) to titanium alloy are shown in Figure 8.4 [25]. The bonding mechanism in this case is believed to be based on the fact that the Ti in Ti—Ni solder selectively accumulates on the sapphire interface, part of which is oxidized, and the resulting titan oxides (TiO_2, TiO, TiO_3) react with sapphire; or Ti ions diffuse into sapphire crystals to form on the bonded interface an intermediate layer consisting primarily of Ti—O—Al-based solid solutions or compounds, which constitutes a solid and vaccum-tight bonded body.

Based on the surface energy of various elements and the solubility of Ti therein, Nicholas et al. have shown that in addition to Sn, In, which has low surface energy and is easily saturatable with Ti, is a very favorable alloying element, a small amount of which will improve wettability remarkably, as shown in Figure 8.5, and that Al, Au, and Ag are moderately favorable elements, Ga and Ni being nonfavorable or even deteriorating elements [26].

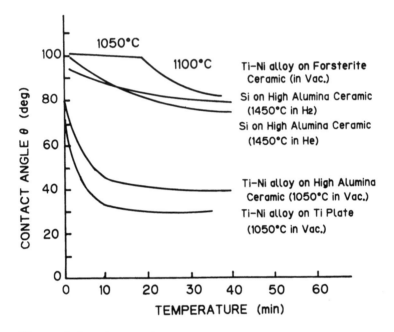

Figure 8.3   Change in contact angle of metal-ceramic with time.
(Ref. 1.)

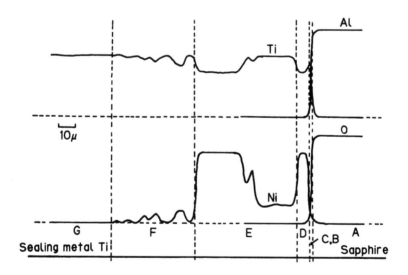

Figure 8.4   Diffusion aspects of elements in the sealed part.   (Ref.
25.)

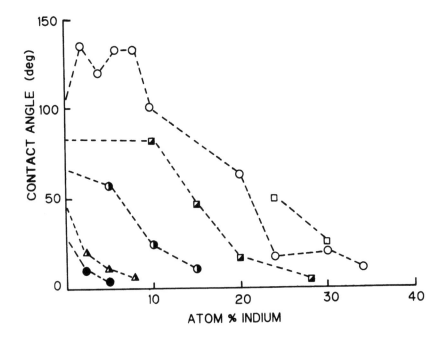

Figure 8.5 Effects of additions of indium on the wetting of alumina by copper-titanium alloys at 1150°C. □, 2.5 Ti; ○, 5 Ti; ▫, 10 Ti; ◐, 15 Ti; ◮ 20 Ti; ●, 25 Ti. (Ref. 26.)

Moreover, although there are many reports dealing with interface products produced using various sorts of ceramic-to-metal bondings effected in vacuum or in the atmosphere by means of various solders, they do not, even when the experiments are similar, always agree [27-30]. Based on experimental results by Economos et al. [31] on typical examples obtained on interfacial reactions between oxides and ceramics at 1800°C in an argon atmosphere, they suggest that there are four types of reactions, as shown in Table 8.3: (1) where new phases are clearly formed on interfaces, (2) where oxide interfaces are attacked by metals, (3) where metals produce penetration along crystal boundaries of oxides, and (4) where changes are barely perceptible on interfaces. Among them, (3) takes place along with reactions of (2) and (3), often observed on oxide-metal systems at high temperatures.

8.3.3 Bonding by a Solid-Phase Reaction Bonding Process

In a solid-phase bonding process, ceramics and metals are processed under pressure at a temperature (about $0.9 \times T_m$) lower than the

Table 8.3 Four Types of Interfacial Reactions of Metals with Oxides at 1800°C in a Neutral Atmosphere

Type				
Formation of a definite new phase at the interface	Be—Al_2O_3, Nb—BeO	Be—MgO, TiO—MgO	Si—Al_2O_3, —	Si—MgO, —
Corrosion of the oxide interface by the metal	Ti—BeO, Ti—MgO	Ti—Al_2O_3, Zr—Al_2O_3	Ti—TiO_2, Zr—TiO_2	Ti—ZrO_2, Zr—MgO
Penetration along the grain boundaries and alteration of the oxide phase	Be—BeO, Be—ThO_2, Nb—BeO, Ti—BeO, Ti—ThO_2, Zr—ThO_2, Mo—Al_2O_3 (2100°C)	Be—Al_2O_3, Si—BeO, Nb—ZrO_2, Ti—Al_2O_3, Zr—BeO, —, —	Be—ZrO_2, Si—Al_2O_3, Nb—MgO, Ti—ZrO_2, Zr—Al_2O_3, —, —	Be—MgO, Si—ThO_2, Nb—ThO_2, Ti⟋MgO, Zr—ZrO_2, —, —
No physical alteration of the metal metal-oxide interface	Ni—BeO, Ni—MgO, Mo—TiO_2, Be—BeO, Nb—TiO_2, Ti—ThO_2	Ni—Al_2O_3, Ni—ThO_2, Mo—ZrO_2, Be—ZrO_2, Nb—ZrO_2, Zr—ZrO_2	Ni—TiO_2, Mo—BeO, Mo—MgO, Be—ThO_2, Nb—MgO, Zr—ThO_2	Ni—ZrO_2, Mo—Al_2O_3, Mo—ThO_2, Nb—Al_2O_3, Nb—ThO_2, —

Source: Ref. 31.

melting point (T_m) of metals—in other words, without being involved in a liquid phase [32].

Particularly when high-temperature construction materials are sought, it is necessary to select a ceramic and a metal whose thermal expansion coefficients are similar, or to arrange an appropriate intermediate layer between the ceramic and the metal to increase thermal shock resistance. The bonding availability of oxide ceramics to films of noble metals such as Pt and Au is based on the fact that these soft metals are chemically inert against environmental conditions [33–35]. Pt presents a thermal expansion coefficient close to those of most oxide ceramics, and the mechanical strength and airtightness of its bonded systems are unchanged through thermal cycles from room temperature to high temperatures. Because of its high Pt content, an Al_2O_3-Pt system (1550°C, 0.03 MPa, 100 s) has a bonding strength of 250 MPa, whereas a stabilized ZrO_2-Pt-Al_2O_3 system (1459°C, 1 MPa, 4 h) has a bonding strength 150 MPa [35].

The eventual corrosion reactions taking place on the interface between ceramics and noble metals were researched by means of thermodynamic values, the results of which are shown in Figure 8.6. It is particularly interesting to note that the Pd-MgO system is involved in a diffusion reaction when kept below 800°C and in a diffusionless reaction when kept above 800°C [36]. A soft metal process in which a low-melting, soft Al is applied as the intermediate layer is also under study. One typical example of bonding is that of Al_2O_3 to stainless steel [2]. For a 1-mm-thick intermediate layer of Al (the thickness of Al after bonding is 0.17 mm) for which bonding was by means of a hot press under vacuum conditions, a pressure of 50 to 100 MPa, and a duration of 30 min, a bonding strength of 40 MPa was obtained, as shown in Figure 8.7. Moreover, a strain-removing annealing effected at 100°C after bonding gives a strength of about 70 MPa.

Studies in which intermediate layers of Al are applied for nonoxide ceramics such as SiC and Si_3N_4 have been reported. For example, SiC bonding by two methods, reaction sintering and the normal pressure sintering process, were researched [37]. Under vacuum at a temperature of 800 to 1000°C for 2 to 3 min, bonding strengths of 240 MPa and 110 MPa, respectively, were obtained for normal pressure sintered SiC and reaction sintered SiC, as shown in Figure 8.8. In addition, on the bonded interface of the former, Al_4C_3 was formed according to the following equation:

$$3SiC_{(s)} + 4Al_{(1)} \longrightarrow Al_4C_{3(s)} + 3Si_{(s)}$$

showing good coordination with SiC. Al_4C_3 shows considerable deterioration in water, however; dipping it in boiling water for a week or so rapidly reduces the bonding strength. When Si_3N_4 was bonded

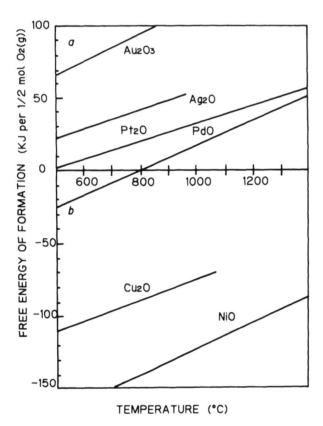

Figure 8.6 Free energies of formation for metal oxides: (a) type 1; (b) type 2. (Ref. 36.)

in an argon atmosphere (1073 K, 0.15 MPa, 7 min), a bonding strength of 4 to 10 MPa was obtained, which is insufficient [38]. Recently, the pressureless bonding of 3Y-TZP (ZrO_2 containing 3 mol% Y_2O_3) and Ti metal, using a bonding agent containing Cu_2O added 5 wt% activated carbon as a reductant at 1000°C, was reported by Kogure et al. From analysis of the interfacial composition, the structure of the bonded interface comprised of plural layers as shown in Figure 8.9. The reaction layers from 3Y-TZP sequential to Ti metal are a titanium oxide layer (TiO, Ti_2O)/Ti rich layer and a (Ti with scanty Cu)/Cu-Ti alloy layer ($CuTi_2$).

Utilizing the same method, the bonding of 3Y-TZP with Fe, Nb and Ta has been examined. With the 0.5 mm Ti sheet, the bonding strength of TZP with Fe, Nb and Ta specimen has become more than 140 MPa, 105 MPa and 120 MPa, respectively [39].

8.3.4 Field-Assisted Bonding

Field-assisted bonding is based on using the ion properties of an element to force other elements to transfer on the interface so as to induce a reaction, making bonding possible even at low temperatures. This process was first tried for sealed adhesion of glass to metals, but has also been tried for β-alumina bonded products [40–41]. Bonding of β--Al$_2$O$_3$ to Cu, Fe, Mo, Ti, Al, and Kovar alloy were effected under an N$_2$ atmosphere at 500 to 600°C, 50 to 500 V, and 0.5 to 10 mA for 45 to 600 min. As shown in Table 8.4, good bonding was obtained on all metals. The formation of metallic oxides was observed on the bonded surfaces.

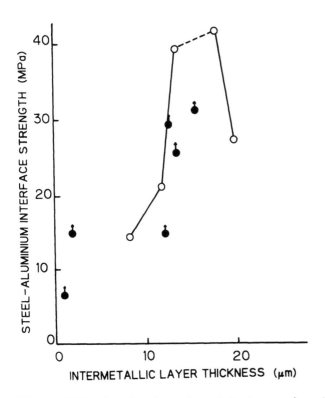

Figure 8.7   Sample strengths plotted as a function of the thickness of intermetallic layers formed at steel-alumina interfaces. Solid symbols indicate that failure occurred at alumina-aluminum interfaces and hence that the steel-aluminum interface strengths were higher. (Ref. 2.)

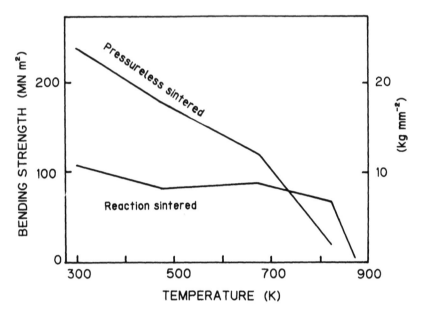

Figure 8.8 Strength in four-point bending at high temperature. Pressureless-sintered and reaction-sintered SiC were joined by liquid aluminum at 1273°C. (Ref. 37.)

8.3.5 Bonding of Non-Oxide-Based Ceramics

Compounds of metallic elements with carbon, nitrogen, boron, silicon, and so on, which present high melting points, high thermal and electric conductivity values, and thermal shock resistance greater than those of oxides are promising as heat-resistant materials. For example, Si_3N_4, BN, AlN, TiN, SiC, ZrC, B_4C, HfC, W_2C and ZrB_2 can all be cited. SiC and Si_3N_4, in particular, are subjects of research and development.

In the case of bonding of reaction-sintered SiC ceramics containing free Si, an example of effective application of the free Si is for bonding realized under vacuum at 1180°C for 10 min, with Ge used as a solder. An interface layer of Si—Ge solid solution was formed and a four-point bending strength of 150 MPa was obtained. Moreover, with the Ge layer thickness brought up to 20 μm, an increase in bonding strength to 400 MPa at near 1000°C was observed [42]. When bonding SiC to stainless steel (316), which have thermal expansion coefficients of 5×10^{-6}°C^{-1} and 18×10^{-6}°C^{-1} respectively, a ductile solder alloy is applied to decrease stress on the interface.

For example, when Ag—Cu—Ti alloy is used, adding 1.5 wt % Ti to a normal pressure-sintered SiC resulted in good wettability and bonding strength. On the interface of these materials a compound

Table 8.4 Bonding Conditions and Results

Metal	Atmosphere	Temperature (°C)	Voltage (V)	Current (mA)	Time (min)	Separation
Cu (preoxidized)	N_2	600	300	2	600	At M—Mo interface
	N_2	550	50	2	120	
Kovar alloy	Air	500	500	10	45	At M—Mo interface
	N_2	500	250	5	600	
Fe	N_2	550	200	2	120	At M—Mo interface
	H_2	550	300	1	120	
Mo	N_2	500	200	2	90	At M—Mo interface
Ti	N_2	600	300	5	45	At M—Mo interface
Al	Air, N_2, H_2	550	300	2	120	Within ceramic
	Air	550	75	0.5	600	

Source: Ref. 40.

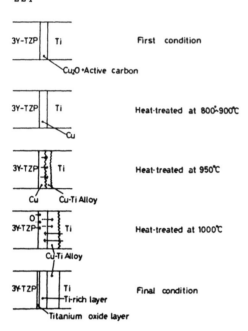

Figure 8.9   Bonding mechanism of Ti/3 Y-TZP system.

phase consisting of Fe and Ti diffused from the stainless steel is
formed together with a reaction layer of SiC with Ti [43]. In the
case of bonding with a reaction-sintered SiC containing free Si,
where Ti forms compounds with the Si, the wettability is hindered
and Ti must be added.

Simultaneously with the diffusion into SiC of Ag and Cu in a
solder alloy, there is diffusion of Si into the solder, and an increase
in intermetallic compounds in Cu—Si and Si—Ti systems embrittles
the interface phase and makes the removal of thermal stress difficult. Direct bonding of SiC is difficult even under high pressure
because of SiC retains a high level of resistance even at temperatures
near 1500°C. Therefore, a finely divided easy-to-sinter SiC was put
in "sandwich" form together with Al, B, and C and subjected to hot
pressing at 1650°C under a pressure of 50 MPa and 30 min, which resulted in a four-point bending strength of 540 MPa [44]. In this
intermediate phase a phase of $Al_8B_4C_7$ is formed. Also, in the bonding of reaction-sintered SiC's among themselves, a powdery mixture
of SiC with C was applied as the intermediate layer and silicified at

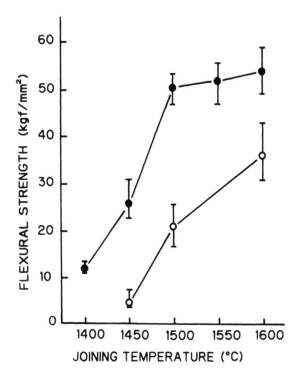

Figure 8.10   Dependence on joining temperature of flexural strength of joined specimens (Si_3N_4) measured by means of the four-point bending method. •, pressure joining, 20 MPa, 30 min; o, pressure-less joining, 30 min. (Ref. 46.)

1450°C under vacuum for 30 min, which resulted in a bonded product with a bonding strength of 520 MPa [45].

For sintered Si_3N_4 bonding, using a hot press at 1400 to 1600°C under 20 MPa for 30 min, a bonded product with a four-point bending strength up to 530 MPa was obtained, which is similar to the matrix strength. Without pressure, a strength of 320 MPa was obtained. Bonding temperature in relation to the strength obtained is shown in Figure 8.10 [46]. Moreover, the bonding of Si_3N_4 without an additive in a range of high bonding pressure up to 3 GPa was researched, and bonding under 3 GPa at 1800°C for an hour gave a bonding structure with an indiscernible bonding interface [47]. For hot press bonding of Si_3N_4, the use of powdery ZrO_2 and $ZrSiO_4$ has been recommended, as the addition of ZrO_2 is so effective that even

under a pressure of less than 1.5 MPa, a bonding strength of 175
MPa can be obtained [48]. Bonding by means of oxide glasses may
be applicable even for complicated configurations and durable at a
wide range of temperatures [49,50].

8.3.6   Bonding of Glass to Metal

Bonding glass to metal is common in the manufacture of enameled
wares and in the sealed adhesion of metallic core wires or caps in
glass vacuum systems. In these cases, where the glass is usually
brought to a molten state and bonded to metal, it is necessary that
the molten glass wet the metal sufficiently and that the thermal expansion coefficients of both phases be well adapted to each other.

As iron oxides are very easy to dissolve in glass, almost no un-
dissolved oxide remains on the interface in the usual bonding conditions. A process involving iron surface oxides dissolving into glass
is shown in Figure 8.11 [51]. Here t_0 is the instant when oxidized
iron comes into contact with the glass, t_1 an instant after the elapse
of sometime, t_2 an instant before the monomolecular layer of the fin-
al oxide dissolves, and t_3 an instant when, after the diffusion of
oxide into the glass, the oxide concentration of the glass layer adjacent to the iron has been lowered considerably below equilibrium.
Of course, considered solely on the interface, the iron and the glass
at t_2 may be regarded as being in equilibrium. For a solid chemical
bond to be formed however, it is necessary that a continuous electronic structure be formed across the interface, which is satisfied
when the glass layer saturates metallic oxide.

In case of bonding $Na_2Si_2O_5$—FeO-based glasses to iron plates
under a total pressure of 10^{-8} atm, an oxygen partial pressure of
$pO_2 < 10^{-20}$ atm, and heating at 1000°C, the contact angle decreases
rapidly from 55° to 10° with increase in FeO from 0 to 6.0 wt %, a
glass expansion taking place at an addition of FeO $\geqslant$ 9.1 wt %. On
the other hand, at an oxygen partial pressure, $pO_2 \sim 10^{-10}$ atm, as
FeO increases from 0 to 44.5 wt %, the contact angle decreases con-
tinuously from 60° to 27° [52]. On the glass/iron plate interface,
oxidation-reduction reaction (1) below takes place, the change in
free energy of reaction being given in equation (2):

$$Fe + Na_2O(glass) \longrightarrow FeO(glass) + 2Na\uparrow \tag{1}$$

$$\Delta G = \Delta G° + \frac{RT \ln[\alpha-FeO \; glass(p_{Na})^2]}{\alpha-Na_2O \; glass} \tag{2}$$

The standard free energy $\Delta G°$ always being positive, due to
the oxidation potential of Na being larger than that of Fe, the

reaction of equation (1) does not proceed under natural conditions. In order that ΔG be negative, it is necessary that the activity term of equation (2) be smaller than unity. An increase in FeO results in an increase in the O/Si ratio of $Na_2Si_2O_5$ glass, which causes a rapid increase in α—Na_2O glass, and as a result, a decrease in ΔG. In addition, due to an atmospheric pO_2 higher than the pO_2 in equilibrium with FeO at 1000°C, $\simeq 2 \times 10^{-5}$ atm), α—FeO metal is almost equal to unity ($\simeq 1$) on the interface even when there is an imperceptible oxide layer, whereas α—FeO glass is less than unity ($\ll 1$). Such a difference in chemical potential causes oxide diffusion to proceed.

When a particularly high bonding strength is required, Fe—Cr alloys or Fe—Ni—Cr alloys are used. Sealing metals oxidized in wet hydrogen gas so as to form a film of Cr_2O_3 or other compound on their surface are bonded to glass. Analytical results through Auger spectroscopy of the diffusion aspects of elements on the interface between an Fe—Cr alloy and glass are shown in Figure 8.12. Based on the fact that the color tone of the bonded portion is entirely the same as that of the surface oxide and that the bonding of the surface oxide to the Fe—Cr alloy is solid, the bonding mechanism is considered to be a thin, dense oxide layer interposed structure [53].

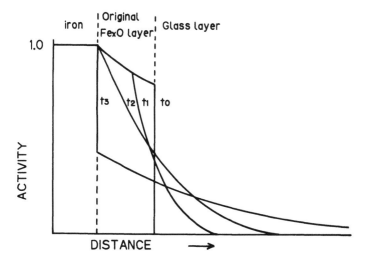

Figure 8.11   Diagram of hypothetical iron activity versus penetration distance for oxidized iron-glass contact zone showing ferrous iron activity in the oxide and the glass relative to metallic iron as the standard state. (Ref. 51.)

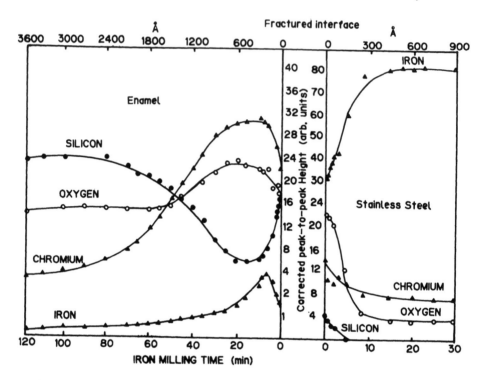

Figure 8.12   Profile analysis of a stainless steel/enamel interface.
(Ref. 53.)

Dumet wire, conventionally used as the enclosed wire in glass
bulbs, is a copper-coated Fe—Ni alloy wire with cuprous oxide formed on its surface, which is protected with borax to prevent the copper from excess oxidation. In one study, when CuO was melted in
glass with mixing, and scaled to copper, forming a Cu_2O layer on
the copper surface, an increase in bonding strength resulted [54].
The diffusion aspects of elements on the bonded area are shown in
Figure 8.13 [55,56]. In any case, such metallic oxide layers play
an important role in glass-metal sealed adhesion.
    Kovar, an Fe—Ni—Co alloy, has very good bonding strength to
boron silicate glass. In this case, the most oxidizable metallic constituent (Fe) is selectively dissolved in the glass; the dissolution of
the other metals (Co and Ni) is considerably inhibited, and as a result they concentrate on the interface. Figure 8.14 shows the diffusion aspects of the elements. The bonding strength is influenced
by the thickness of the intermediate glass layer in which oxides dissolve or by the concentration of oxides, and excess production of

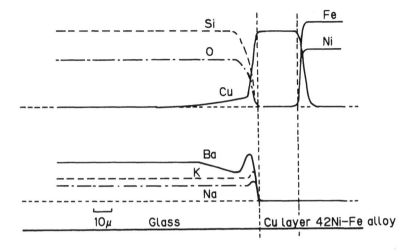

Figure 8.13 Diffusion aspects of elements in the sealed part of sample 2. (Ref. 55.)

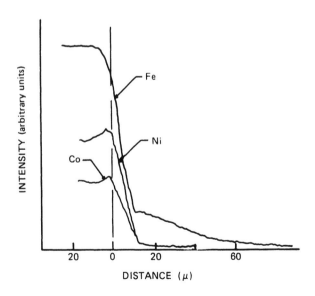

Figure 8.14 Diffusion aspects of elements in a glass-Kovar bonding interface. (Ref. 57.)

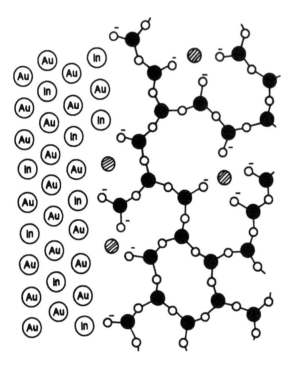

Figure 8.15   Sketch of structure near the interface between an
Au—In alloy and silica glass. Closed, open, and hatched circles
represent silicon atoms, oxygen atoms, and indium ions. (Ref. 58.)

oxides such as FeO and Fe_2O_3 in the course of preliminary oxidation
will result in devitrification on the bonded interface [57].

The bonding of Au—In alloys to silicate glasses presented a remarkably improved bonding strength when In was vacuum-evaporated
to a silica glass surface in advance and heated in air to form a In_2O_3
film [58]. Oxides heated at high temperature dissolve in the silica
glass and destroy the tetrahedral net structure of SiO_4, generating
nonbridging oxygen. This causes a dissolution of In ions to take
place, which requires oxygen ions as follows:

$$\equiv Si-O-Si\equiv \; + \; In^{3+} + O^{2-} \longrightarrow \; \equiv Si-O^-In^{m+}O^--Si\equiv$$

$$(3)$$

Thus, while the reaction taking place between In oxide layer
and silica glass produces a lot of nonbridging oxygen on the interface, the strongest bonding is obtainable when the oxide layer

dissolves completely in the silica glass and disappears, upon direct combination of metal element with nonbridging oxygen, as shown in Figure 8.15.

REFERENCES

1. H. Takashio, Bonding of Ceramics to Metal, *Surf. Sci. Jpn.*, *4*, 2 (1983).

2. M. G. Nicholas and R. M. Crispin, Diffusion Bonding Stainless Steel to Alumina Using Aluminium Interlayers, *J. Mater. Sci.*, *17*, 3347 (1982).

3. T. Okuo, Deterioration of Ceramics-Metal Bonded Structures Under High Heat Flux Conditions, *J. High Temp. Soc. Jpn.*, *6*, 200 (1983).

4. O. Nomura, Y. Ebata, and K. Hijikata, Thermal Resistance of Buffer Layer in a Ceramic Wall of MHD Generation Channel, *J. Ceram. Soc. Jpn.*, *90*, 658 (1982).

5. K. Suganuma, T. Okamoto, M. Koizumi, and M. Shimada, New Method for Solid-State Bonding Between Ceramics and Metals, *J. Am. Ceram. Soc.*, *66*, C-117 (1983).

6. K. Suganuma, T. Okamoto, M. Shimada, and M. Koizumi, Solid-State Bonding of Oxide Ceramic to Steel, *J. Nucl. Mater.*, *133 & 134*, 773 (1985).

7. S. Morozumi, M. Kikuchi, S. Sugai, and M. Hayashi, Reaction Between Molybdenum and Carbon, and Several Carbides, *J. Jpn. Inst. Met.*, *44*, 1404 (1980).

8. S. Morozumi, M. Kikuchi, and S. Sugai, Reaction Between Molybdenum and Several Nitrides, *J. Hpn. Inst. Met.*, *45*, 184 (1981).

9. S. Morozumi, M. Kikuchi, and T. Nishino, Bonding Mechanism Between Alumina and Niobium, *J. Mater. Sci.*, *16*, 2137 (1981).

10. R. F. Pabst and G. Elssner, Adherence Properties of Metal-to-Ceramic Joints, *J. Mater. Sci.*, *15*, 188 (1980).

11. S. Morozumi, M. Kikuchi, K. Saito, and S. Mukaiyama, Bonding of Nb to ZrO_2 and HfO_2, *Inst. Met. Prepr. Jpn.*, *94*, 78 (1984).

12. J. T. Klomp, Bonding of Metals to Ceramics and Glasses, *Am. Ceram. Soc. Bull.*, *51*, 683 (1972).

13. A. G. Pincus, Mechanism of Ceramic-to-Metal Adherence, *Ceram. Age*, *3*, 16 (1954).

14. H. Takashio, Alumina Ceramic-to-Metal Seals by the "Mo—Mn Process," *J. Ceram. Soc. Jpn.*, *79*, 330 (1971).

15. E. P. Denton and H. Rawson, The Metallizing of High—Al_2O_3 Ceramics, *Trans. Br. Ceram. Soc.*, *59*, 25 (1960).

16. R. M. Fulrath and E. L. Hollar, Manganese Glass-Molybdenum Metallizing Ceramics, *Bull. Am. Ceram. Soc.*, *47*, 493 (1968).

17. J. T. Klomp, Interfacial Reactions Between Metals and Oxides During Sealing, *Bull. Am. Ceram. Soc.*, *59*, 794 (1980).

18. M. E. Twentyman and P. Popper, High-Temperature Metallizing, *J. Mater. Sci.*, *10*, 791 (1975).

19. L. Leed and R. A. Huggins, Electron Probe Microanalysis of Ceramic-to-Metal Seals, *J. Am. Ceram. Soc.*, *48*, 421 (1965).

20. I. I. Mechelkin, M. A. Pavlova, and A. A. Doronkina, Welding of a High-Alumina Ceramic Material with a Metal, *Svar. Proizvod.*, *7*, 33 (1975).

21. L. C. de Jonghe, H. Schmid, and M. Chang, Interreaction Between Al_2O_3 and a CaO—Al_2O_3 Melt, *J. Am. Ceram. Soc.*, *67*. 27 (1984).

22. W. D. Kingery, E. Niki, and M. D. Nurashimhan, Sintering of Oxide and Carbide-Metal Compositions in Presence of a Liquid Phase, *J. Am. Ceram. Soc.*, *44*, 29 (1961).

23. J. T. Klomp and Th. P. J. Botden, Sealing Pure Alumina Ceramics to Metals, *Bull. Am. Ceram. Soc.*, *49*, 204 (1970).

24. N. Nakahashi, M. Shirokane, and H. Takeda, Bonding of Nitrate Ceramic to Metal, *Surface*, *24*, 595 (1986).

25. H. Takashio, Sapphire to Metal Seal by Reactive Alloying Method, *J. Ceram. Soc. Jpn.*, *78*, 350 (1970).

26. M. G. Nicholas, T. M. Valentine, and M. J. Waite, The Wetting of Alumina by Copper Alloyed with Titanium and Other Elements, *J. Mater. Sci.*, *15*, 2197 (1980).

27. H. J. De Bruin, A. F. Moodie, and C. E. Warble, Ceramic-Metal Reaction Welding, *J. Mater. Sci.*, *7*, 909 (1972).

28. A. J. Mceroy, R. H. Williams, and I. G. Higginbotham, Metal/Non-metal Interfaces: The Wetting of Magnesium Oxide by Aluminium and Other Metals, *J. Mater. Sci.*, *11*, 297 (1976).

29. C. A. Calow, P. P. Bayer, and I. T. Porter, The Solid State Bonding of Nickel, Chromium and Nichrome Sheets to α—Al_2O_3, *J. Mater. Sci.*, *6*, 150 (1971).

30. J. J. Brennan and J. A. Pask, Effect of Nature of Surfaces on Wetting of Sapphire by Liquid Aluminum, *J. Am. Ceram. Soc.*, *51*, 569 (1968).

31. G. Economos and W. D. Kingery, Metal-Ceramic Interactions: II. Metal-Oxide Interfacial Reactions at Elevated Temperatures, *J. Am. Ceram. Soc.*, *36*, 403 (1953).

32. F. P. Bailey and K. J. T. Black, Gold-to-Alumina Solid State Reaction Bonding, *J. Mater. Sci.*, *13*, 1045 (1978).

33. R. V. Allen, Solid State Metal-Ceramic Reaction Bonding Applications to Transistor Packages and Advanced Materials, *Electrocomponent Sci. Technol.*, *11*, 85 (1983).

34. R. V. Allen and W. E. Borbidge, Solid State Metal-Ceramic Bonding of Platium to Alumina, *J. Mater. Sci.*, *18*, 2835 (1983).

35. R. V. Allen, W. E. Borbidge, and P. T. Whelan, Advances in Ceramics, 12, 537, (1984).

36. H. J. Bruin, Interfacial Noble-Metal Corrosion in Metal to Ceramic Reaction Welding, *Nature*, *272*, 712 (1978).

37. T. Iseki, T. Kameda, and T. Maruyama, Interfacial Reactions Between SiC and Aluminium During Joining, *J. Mater. Sci.*, *19*, 1692 (1984).

38. K. Suganuma, T. Okamoto, and M. Koizumi, Method for Preventing Thermal Expansion Mismatch Effect in Ceramic-Metal Joining, *J. Mater. Sci. Lett.*, *4*, 648 (1985).

39a. E. Kogure, H. Hoshino, T. Iida, and T. Mitamura, Pressureless Bonding of 3Y-TZP to Titanium, *J. Ceram. Soc. Jpn.* *96*, 1057 (1988).

39b. E. Kogure, H. Hoshino, T. Iida, and T. Mitamura, Pressureless Bonding of 3Y-TZP and Fe using Ti sheet, *J. Ceram. Soc. Jpn.* *96*, 1142 (1988).

40. B. Bunn, Field Assisted Bonding of Beta-Alumina to Metals, *J. Am. Ceram. Soc.*, *62*, 545 (1979).

41a. G. Wallis and D. I. Pomerantz, Field Assisted Glass-Metal Sealing, *J. Appl. Phys.*, *40*, 3946 (1969).

41b. M. P. Borom, Electron-Microprobe Study of Field-Assisted Bonding of Glasses to Metals, *J. Am. Ceram. Soc.*, *56*, 254 (1973).

42. T. Iseki, K. Yamashita, and H. Suzuki, Joining of Self Bonded SiC containing Free Si by Ge Metal, *J. Ceram. Soc. Jpn.*, *89*, 171 (1981).

43. T. Iseki, H. Matsuzaki, and J. K. Boadi, Brazing of Silicon Carbide to Stainless Steel, *Am. Ceram. Soc. Bull.*, *64*, 322 (1985).

44. T. Iseki, K. Yamashita, and H. Suzuki, Joining of Dense Silicon Carbide by Aluminum Metal, *J. Ceram. Soc. Jpn.*, *91*, 11 (1983).

45. T. Iseki, M. Imai, and H. Suzuki, Joining of Dense Silicon Carbide Containing Free Silicon by Reaction-Sintering, *J. Ceram. Soc. Jpn.*, *91*, 259 (1983).

46. S. Kanzaki and H. Tabata, Diffusion Joining of Silicon Nitride Ceramics, *J. Ceram. Soc. Jpn.*, *91*, 520 (1983).

47. T. Kaba, M. Shimada, and M. Koizumi, Diffusional Reaction-Bonding of Si_3N_4 Ceramics Under High Pressure, *Commun. Am. Ceram. Soc.*, *66*, C-135 (1983).

48. P. F. Becher and S. A. Halen, Solid State Bonding of Si_3N_4, *Am. Ceram. Soc. Bull.*, *58*, 582 (1979).

49. R. D. Brittain, S. M. Johnson, R. H. Lamoreaux, and D. J. Rowcliffe, High-Temperature Chemical Phenomena Affecting Silicon Nitride Joints, *J. Am. Ceram. Soc.*, 67, 522 (1984).

50. S. M. Johnson and D. J. Rowcliffe, Mechanical Properties of Joined Silicon Nitride, *J. Am. Ceram. Soc.*, *68*, 468 (1985).

51. M. P. Borom and J. A. Pask, Role of Adherence Oxides in the Development of Chemical Bonding at Glass-Metal Interfaces, *J. Am. Ceram. Soc.*, *49*, 1 (1966).

52. C. E. Hoge, J. J. Brennan, and J. A. Pask, Interfacial Reactions and Wetting Behavior of Glass-Iron Systems, *J. Am. Ceram. Soc.*, *56*, 51 (1973).

53. D. E. Clark, C. G. Pantano, Jr., and G. Y. Onoda, Jr., Auger Analysis of Brass-Enamel and Stainless Steel-Enamel Interfaces, *J. Am. Ceram. Soc.*, *58*, 336 (1975).

54. Y. Sameshima, Chemical Bonding of Metal to Glass for Airtight Seals, *J. Chem. Soc. Jpn. Ind. Chem. Sect.*, *73*, 35 (1970).

55. H. Takashio, Alkali Barium Silica Glass-to-Dumet Wire Seals, *J. Ceram. Soc. Jpn.*, *83*, 315 (1975).

56. H. Takashio, Dumet-to-Alkali Lead Silicate Glass Seals, *J. Ceram. Soc. Jpn.*, *84*, 383 (1976).

57. Y. Ikeda, Mechanism on Glass to Metal Bonding, *J. Soc. Mater. Sci. Jpn.*, *17*, 783 (1968).

58. H. Ohno, T. Ichikawa, N. Shiokawa, S. Ino, and H. Iwasaki, ESCA Study on the Mechanism of Adherence of Metal to Silica Glass, *J. Mater. Sci.*, *16*, 1381 (1981).

9

Interfacial Modifications and Bonding of Fiber-Reinforced Metal Composite Material

9.1   INTRODUCTION

Fiber-reinforced metal composite materials are composed of metals combined with inorganic, metallic, or ceramic fibers whose properties are different from those of the metals, so that the metal and the fibers have a maximum effect on their respective properties and compensate for each other's shortcomings. The interfaces between these combinations of materials play a very important role. It would not be an exaggeration to say that the properties of the composite material are determined by the interfacial structure. The roles of the interface in composite materials are generally divided into two groups: (1) dynamic properties on the interface, and (2) physicochemical compatibility.

The dynamic properties are related to the reinforcement mechanism in the composite material, which influences failure phenomena based on differences in the heat expansion coefficients, Poisson's numbers, and other characteristics of the two materials in the composite. Physicochemical compatibility relates to adhesion, penetration, mass transfer, diffusion, reaction, and other parameters that interact between the combined materials. In most cases, fiber-reinforced metal materials are produced at high temperature according to a solid- or liquid-phase processing method, which is different from that for polymer-type composite materials. Various processing methods for fiber-reinforced metal composite materials have been proposed [1] as shown in Figure 9.1.

In brief, solid-phase processing methods are those in which reinforcing fibers and a matrix metal are pressed together at a temperature below the melting point of the metal, and solidified integrally. On the other hand, liquid-phase processing methods are those in which a molten metal is cast into the area surrounding the fibers at

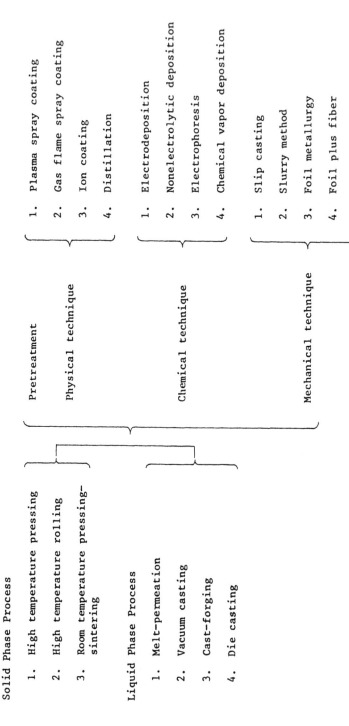

Figure 9.1   Various processing methods for fiber-reinforced metal composite materials.   (Ref. 1.)

a temperature above the melting point of the matrix metal, and cooled for solidification. Needless to say, in liquid-phase processing methods, reinforcing fibers and a matrix metal are subjected to relatively high temperatures.

When such diffusion, reactions, and so on arise on the interface, a new phase is formed. More particularly, the new phase is formed by reactions, diffusion, and dissolution between the fibers and the matrix. As a result, fibers lose their reinforcing effect, and in some cases, the composite material, when measured, shows lower properties than those of the matrix metal only.

Also, in solid-phase processing methods, there is no way to avoid the fact that a matrix metal and reinforcing fibers must be subjected to high temperatures to facilitate plastic deformation of the matrix metal and sufficient bonding of the boundary faces between the fibers and the metal and between the matrix metals. Accordingly, the ideal interface takes the form of physical bonding in the morphology. However, thermodynamically, the interface is in the region where some thermal vibration may cause diffusion, reactions, and so on.

For an evaluation of the strengths of a composite material, one method utilizes a rule of mixture in which values obtained independently by measurement of the matrix and reinforcing fiber are used. Therefore, the rule of mixture is not adaptable for those cases where a new phase is formed. As should be clear from the description above, the interfaces in composite materials are critical parameters that require further consideration. Extensive studies have been undertaken on the surface treatment of reinforcing fibers, the modification of matrix metals by the addition of alloy elements, and so on, with the goal of completely avoiding the formation of a new phase on the interface, and also to enhance the compatibility of the materials being combined. In this chapter we discuss interfacial modifications and bonding of fiber-reinforced metal composite materials that utilize typical fibers.

9.2 CARBON-FIBER-REINFORCED METAL COMPOSITES

Carbon fibers have high strength and modulus of elasticity, are lightweight, and are relatively inexpensive. Moreover, there are so many types of carbon fibers that a most appropriate carbon fiber can be selected for use in a composite material. As a reinforcing fiber for fiber-reinforced metal (FRM), use of carbon fibers has become more popular. However, they have great shortcomings, such as high reactivity with metals and poor wettability with molten metals. A typical lightweight, high-strength matrix metal for FRM is aluminum (Al). With regard to the reactivity of carbon fibers with Al, it has been reported that PAN (polyacrylonitrile)-type high-strength carbon fibers have the greatest

tendency to react with Al, followed by PAN-type high-elastic carbon
fibers, and pitch-type carbon fibers, in decreasing order of reactivity
[2].

As hown in Table 9.1 [3a], in the case of various composite materials utilizing different carbon fibers and Al, Al_4C_3, an interphase
compound, is produced on the interfaces according to the following
reaction formula:

$$3C + 4Al \longrightarrow Al_4C_3$$

From recent research, the degradation of PAN-based carbon fibers depends on the temperature and holding time of molten aluminum. At
lower temperature, the high-strength type carbon fibers were more
degraded than the high-modulus type carbon fibers in the early stage,
but the high-modulus type carbon fiber were more degraded in the
later stage. At higher temperatures, the high-modulus type carbon
fibers were more degraded than the high-strength type carbon fibers
from the early stage. The difference in the degradation behavior of the
carbon fibers is attributable to the morphology of Al_4C_3 crystals formed
at the interface between the fiber and molten aluminum [3b]. Such formation of interphase compounds results in weakening of the mechanical
properties of the composite materials. Failure results from cracks in
the coating layer. Accordingly, various attempts have been made to
develop surface treatments for carbon fibers. To enhance the compatibility and wettability of carbon fibers with Al, the surfaces of carbon
fibers are coated with SiC [4], TiC [5a], Ti—B [5b], CrC_2 [6], or a
similar compound as an intermediate layer. It has been reported that
by coating the fibrous surfaces with an intermediate layer of SiC, TiC,
or TiB according to the CVD method, the surface-coated fibers can be
highly stabilized in molten metal, resulting in improved compatability
and wettability.

An intermediate layer of CrC_2 was obtained by electrodeposition of
Cu and Cr into double layers, with heat treatment for the formation of
CrC_2. Although the CrC_2 intermediate layer is broken in molten aluminium, failure can be inhibited by the addition of Mg to the Al and reduction of the activity of the Al. Such ceramic intermediate layers are
designed to enhance the compatibility of the carbon fibers by utilization
of a reaction inhibiting effect. On the other hand, to improve the wettability of carbon fibers, it is effective to coat the fibers with a metal
or a metal alloy to be used as the matrix of the composite material. For
this purpose, there is a method (1) wherein the fibers are passed
through a molten metal generally for use as the matrix of the composite
material.

In many cases the fibers are previously surface-treated to improve
wettability with the molten metal. For example, the fibers are sequentially dipped in molten metal liquids of Na, Sn, and Al or an Al alloy [7].
The Na molten metal liquid functions as a wetting agent for carbon fibers,

Table 9.1 Degradation of Various Graphite Fibers After Titanium-Boron Treatment and Infiltration with Aluminum Alloys

Type	Fiber	Fiber degradation after TiB treatment (%)	Al_4C_3 formation (ppm)	Composite degradation from Al_2O_2 infiltration (%)	Total composite strength degradation (%)
Rayon	Thornel 50	8	250	2	9
	Thornel 75	—	250	—	6
Pitch	Type P	—	100	—	22
Pan I	Modmor I[a]	10	125	9	18
Pan II	Thornel 300	32	> 6000	39	72
	Hercules A	—	> 6000	—	78

[a]Not surface-treated.
Source: Ref. 3a.

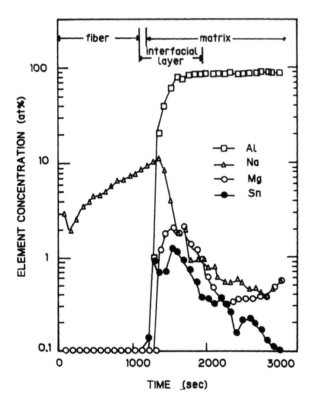

Figure 9.2 IMMA plot of sodium, tin, magnesium, and aluminum concentrations as a function of time (depth of sputtered hole) in graphite-aluminum composite prepared by the sodium process. (Analysis starts in graphite fiber and proceeds into the aluminum matrix.) (Ref. 7.)

and reacts with Sn to form a stable, protective coating layer of an intermetallic compound on the interfaces. Al is substituted for the intermetallic compound, retaining the coating parts on the fibers. Goddard [7] reports adding a small amount of Mg to an Na or Sn bath for high efficiency of Al coating. Figure 9.2 illustrates a distribution of the elements on the interface between a graphite fiber and an Al matrix. Na and Sn exhibit maximum concentrations at the interface, and Mg, which is one component of the matrix, has a relatively high concentration at the interface. While the metal phase is alloyed on the interface, all the metal components except Na are only barely diffused into the fiber structure. It is suggested that Na is diffused among the particles or is present in the intercalcation compounds, such as $C_{64}Na$, formed by reaction with the graphite.

However, there was no evidence that the fibers are eroded. Sn and Na form a series of intermatallic compounds and are melted and mixed with each other completely at the temperature (600°C) of the Sn bath. On the contrary, Na and Al are only difficultly miscible with each other. This is also true of Mg and Na, but Mg and Sn form intermetallic compounds. Accordingly, the carbon fibers are coated with Na—Sn intermetallic compounds, and then with Mg—Sn intermetallic compounds. Table 9.2 lists the mechanical properties of the FRM with carbon fibers treated as described above. Judging from the fact that the values obtained are almost equal to the theoretical values, it is suggested that the fibers adhere sufficiently well to the matrix.

As a second method for coating fibers with a metal or a metal alloy, CVD methods such as ion plating, sputtering, and vacuum deposition are utilized. By these methods, Al, although it has poor wettability with the carbon fibers, can be coated on the fibers. However, it is known that in the case of FRM with Al-coated carbon fibers according to the vacuum, high-temperature pressing method, the tensile strength can be retained up to 450°C, but the FRM deteriorates rapidly in the temperature region 550°C and higher [8,9]. Figure 9.3 shows the test results.

As a third method of coating carbon fibers, electrodeposition, an electrochemical method utilizing a reducing agent for deposition from a metal salt aqueous solution, is exemplified. More particularly, carbon fibers heated to 700°C under vacuum for the activation are dipped in a bath of a metal salt of Cu, Co, Ni, or the like. To the solution is added a substituting agent containing a compound of Mg, Al, Zn, Fe, or the like to form the coating layer, which enhances the wettability of fibers with Al [10]. Also changing the structure of carbon fibers coated with Ni by electroless plating, electrodeposition, vapor deposition, and so on, into that of crystallized graphite is accelerated by heat treatment. Inclinations toward recrystallization decrease in the sequence rayon-type high-modulus fibers, PAN-type high-modulus fibers, PAN-type high-tension fibers.

It has been reported that in the composition of Al with such Ni-coated carbon fibers, the Ni reacts with the Al to form brittle intermetallic compounds within the matrix of the composite material [11]. On the other hand, attempts have been made to add alloy elements to Al matrices, in addition to the surface treatment of carbon fibers mentioned above. Figure 9.4 illustrates changes in the surface tension of Al added with various metals [12]. In general, pure metals have notably high surface tension. However, the surface tension is reduced remarkably by the addition of an alloy element. When the surface tension of a molten metal is lowered, the metal has a higher degree of wettability with the fibers. For example, Al—5 wt % In, Al—1 wt % Pb, and Al—1 wt % Tl exhibit proper wettability with carbon fibers. In particular, the mechanical strength of Al—1 wt % Tl alloy-coated carbon fibers is scarcely reduced. Figure 9.5 shows

Table 9.2   Room Temperature Tensile Strengths of Sodium Process
Aluminum Composites

Composite type (fiber/matrix)	Fiber type	Fiber content (vol %)	Composite fracture stress (MPa)	Composite strength (% theoretical[a])
HM-3000/	Pan I	9.3	210	87.7
1100 Al	Graphite	–	–	–
Thornel 50/	Rayon	13.3	317	96.9
1100 Al	Graphite	–	–	–
Thornel 300/	Pan II	11.3	98	31.8
1100 Al	Graphite	–	–	–
FP/1100 Al	Alumina	13.5	173	117.0
Thornel 50/	Rayon	28.0	731	105.0
A413 Al	Graphite	–	–	–

[a]These values were determined from the equation

$$\sigma_c = \sigma_f V_f + \sigma_m (1 - V_f)$$

where

σ_c = theoretical strength of the composite

σ_f = tensile strength of the fibers, as determined from the strength of an epoxy-infiltrated fiber bundle

V_f = Volume fraction of fibers

σ_m = Stress on the matrix at the breaking strain of the fiber; this value was assumed to be 34.5 MPa for 1100 Al and 207 MPa for A413 Al

Source: Ref. 7.

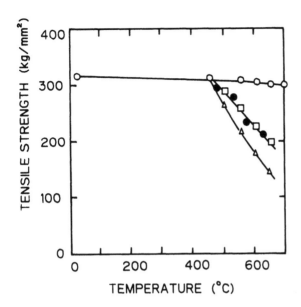

Figure 9.3 Tensile strength of carbon fiber after heat treatment for 1 h under 1×10^{-4} Torr. $\circ$, uncoated; $\triangle$, ion plating; $\square$, vacuum coating; $\bullet$, sputtering. (Ref. 9.)

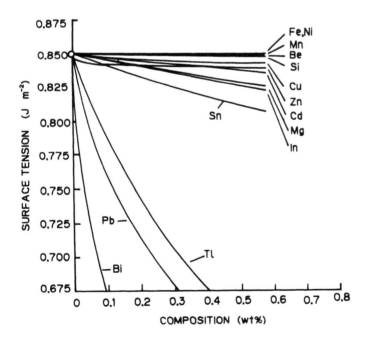

Figure 9.4 Surface tension of aluminum at 973 K with various alloy additions. (Ref. 12.)

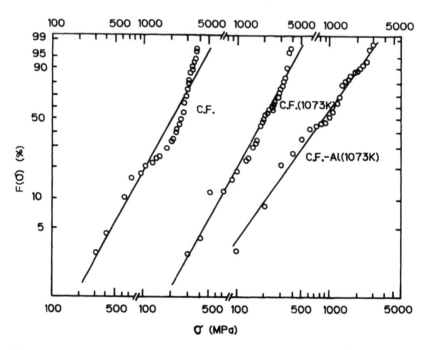

Figure 9.5 Weibull distributions of the tensile strength of as-received carbon fibers (no coating) with or without heating at 1073 K for 1.8 ks and that of pure aluminum-coated fibers after heat treatment. (Ref. 13.)

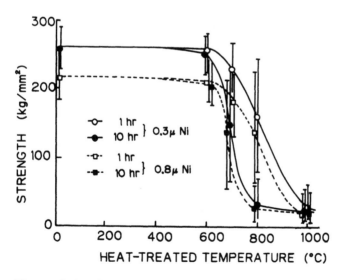

Figure 9.6 Strength of heat-treated carbon fibers coated with nickel. (Ref. 15.)

the test results [13]. It has also been reported that alloys with
Mg, Cu, and Fe have improved compatabilities and are effective for
application to carbon fibers [14].

In addition to the foregoing studies on light-metal-base composite
materials, heat-resisting metal-base composite materials have been investigated extensively, utilizing Ni, Co, Mn, Cr, and the like as
the heat-resisting material. In particular, many studies on Ni-base
composite materials have been carried out. It has been reported
that the Ni in the material causes the carbon fibers to change into
graphite, resulting in deterioration. As shown in Figure 9.6, the
strength of the composite materials decreases remarkably in the temperature range above 600°C [15]. When 20% Cu is added to Ni for
inhibition of the deterioration reaction, the reactivity of the Cu-
added Ni is reduced to about one-tenth that of pure Ni, so that the
graphitization is effectively inhibited [16]. In another study, coated
with chromium carbide (CrC_2) for use as an intermediate layer in
Ni-matrix composite materails, similar to the use of carbon fibers in
Al-matrix composite materials discussed above, were examined. The
report states that Hastelloy-D, which is a Ni-base alloy, has a higher compatibility than pure Ni [6].

9.3 SILICON CARBIDE—FIBER-REINFORCED METAL COMPOSITES

Silicon carbide fibers and filaments have high melting points and are
relatively chemically inactive to air and metals. However, for use
as high-temperature structural materials, it is necessary to improve
the stability of silicon carbide fibers to metals on interfaces. Silicon
carbide fibers are generally divided into two groups: fibers produced by the melt spinning of polycarbosilane and fired by heating,
and fibers obtained by deposition of silicon carbide (SiC) on filaments
made of tungsten and the like according to the chemical vapor deposition (CVD) method.

The polycarbosilane (an organic silicon composite) fiber is composed primarily of noncrystalline silicon carbide and has a smooth
surface without surface faults. Under vacuum, no reduction in the
mechanical strength of the fibers is recognized in the temperature
range up to about 1400°C. However, in the atmosphere, the carbon
contained in the fibers in excess of the stoichiometric composition is
combusted, or SiO_2 is formed, which causes faults in the fiber. Accordingly, the strength of the fiber begins to decrease at about
1000°C, as shown in Figure 9.7 [17]. It has been reported that
polycarbosilane fiber has a higher level of compatibility with relatively more types of metals, as shown in Table 9.3 [18]. In polycarbosilane
fiber/Al composite systems, aluminium carbide (Al_4C_3) begins to be

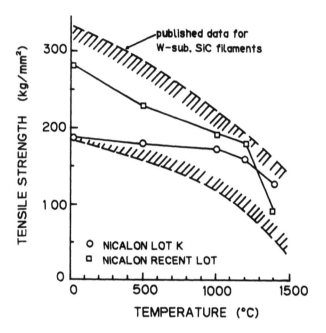

Figure 9.7   Variation of tensile strength in air for Nicalon and
W-substrate CVDd SiC filaments at elevated temperatures.   (Ref.
18.)

formed on the interfaces at 650°C, near the melting point of Al.
As reported previously, Zr has been arc-melted on silicon carbide
fibers, with an ethylene gas (C_2H_4 gas) introduced for the reaction.
This produced ZrC which was vapor-deposited on the surface of
the fibers. Al was also vapor-deposited on the surfaces. It is
reported that the resultant SiC—ZrC—Al system fiber-reinforced
composites, even when heated to 750°C, maintain stable surfaces
[19].

Composite materails of Al combined with silicon carbide fibers
produced by the CVD method suffer less deterioration in strength
in the high-temperature range, as shown in Figure 9.7. However,
the problem of poor wettability on the interface has yet to be solved.
It has been reported that adding Mg, Ni, or Ti to molten matrix [20],
or passing the fibers through molten Al to form a thin coat of Al on
the fiber surfaces [21], is effective in improving poor wettability.
However, this is often accompanied by reaction deterioration of the

Table 9.3 Powder Properties and Reactivity of the Filament with Various Metals

	Powder properties		Sintering temperature[a] (°C)											
	Purity (wt %)	Size (mesh)	450	530	620	650	750	800	850	900	950	1000	1100	1200
Al	99	−200	O	O	O	—	—	—	—	—	—	—	—	—
Ag	>99.9	350	—	—	—	O	O	O	O	O	—	—	—	—
Cu	99.5	150	—	—	—	—	O	O	O	O	O	O	—	—
Ni	99.9	−200	—	—	—	—	—	O	—	O	—	O	×	×
Co	99.8	1–2 μm	—	—	—	—	—	O	—	O	—	△	△	×
Fe	99.9	200	—	—	—	—	—	O	—	O	—	O	O	△
Ti	99	−350	—	—	—	—	—	△	—	△	—	△	△	△
Cr	99.9	−200	—	—	—	—	—	—	—	O	—	O	△	△
Mo	99.9	3.5 μm	—	—	—	—	—	O	—	O	—	O	O	O

[a] O, no reaction was observed on the cross-section of the specimen under the optical microscope (×1000); △, reaction layer and/or edge roughening was observed under the same conditions; ×, filament vanished.
Source: Ref. 18.

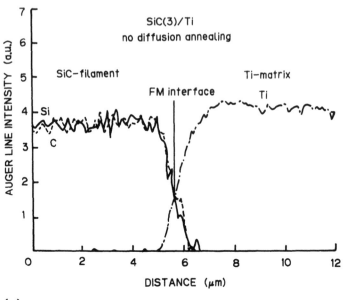

(a)

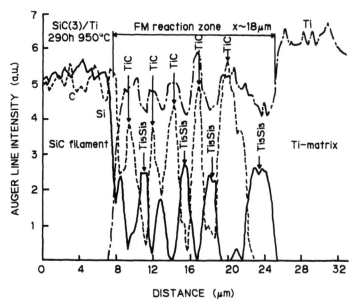

(b)

Figure 9.8   Auger line scans on cross sections of SiC/Ti composites: (a) SiC(3)/Ti as-produced (no diffusion annealing); (b) SiC(3)/Ti annealed 290 h at 950°C.   (Ref. 22.)

fibers. It has been reported that providing intermediate layers of B_4C, TiC, TiN, and the like is effective in solving such problems [20]. In the case of composite materials composed of silicon carbide fibers and titanium rather than Al, multilaminar reaction products of TiC, $TiSi_3$, Ti_5Si_3, and the like are formed on the interfaces, as shown in Figure 9.8 [22]. Addition to the Ti of Mo, Al, Zr, and so on, effectively inhibits the reactions and enhances compatibility [23].

Heat-resistant alloy matrix composite materials, most of which utilize a Ni-base alloy, have also been investigated. In this type of composite material, complicated reactions such as diffusion of Si and C into the metal to form silicides and carbides, and conversely, diffusion of the alloy elements into the ceramics, occur [24]. It has been suggested that an intermediate layer is not effective as a barrier to inhibit the reactions [25].

9.4 BORON-FIBER-REINFORCED METAL COMPOSITES

Boron fibers are produced similarly to chemical vapor deposition for silicon carbide fibers, and the appearance and size of the fibers are analogous to those of silicon carbide fibers. In carbon-core boron fibers, no interface reactions occur. On the contrary, in tungsten-core boron fibers, boride layers of W_2B_5, WB_4, and the like form on the interfaces. Because of its irregular structure, boron reacts vigorously with almost all metals except Ag, Au, Sn, and Be, to form a brittle reaction-product layer. It is known that B and Al, both of which belong to group IIIB of the periodic table, react easily with each other to form AlB_2.

When an Al—Mg alloy is used, AlB_{12} is formed [26]. It is reported that AlB_2 begins to form on the interfaces at the side of the matrix, and AlB_{12} on the interfaces at the side of the fibers, respectively, in the initial period of the reaction. Then the interface between both layers moves toward the surface of the fiber, because of Al's higher diffusion coefficient to pass through the AlB_2 layer than that of B to pass through the AlB_{12} layer. As a result, the AlB_{12} phase disappears.

For the purpose of inhibition of these reactions, intermediate layers of silicon carbide, boron carbide (B_4C), and the like were provided, but the intermediate layers were ineffective for Al and Ti matrix [27]. Test Results have revealed that addition of titanium, zirconium, and hafnium to Al matrixes has a considerable effect [14].

9.5   ALUMINA (Al₂O₃)-FIBER-REINFORCED METAL COMPOSITES

Composite materials of Mg, Al, Pb, Cu, and Zn combined with alumina continuous fibers were produced by the vacuum casting method. Test results have revealed that Mg used as a simple substance has high wettability to alumina fibers, and that the other metals, when add with the alloy elements, exhibit sufficient wettability [28]. For example, it is reported that addition of a small amount (about 3 wt %) of lithium to Al enhances the wettability with the fibers and improves deterioration of the fibers. As shown in Figure 9.9, for alumina fiber/Ni composition, addition of titanium or chromium (Cr) to the Ni considerably reduces the contact angle to the fiber [29]. When Ti is added, α—Ti_2O_3 having a lattice structure analogous to that of alumina is formed. When Cr is added, it is diffused into the

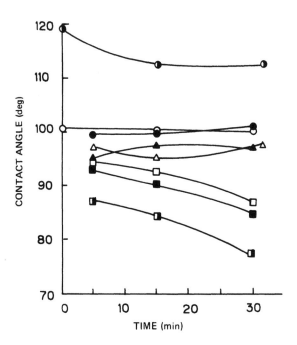

Figure 9.9   Wettability of Ni matrix alloys to Al_2O_3.   ◑, high pure (15ppm); ○, pure (31ppm); ●, Ni 0.82 at % Al; △, Ni 0.71 at % Cu; ▲, Ni 0.76 at % In; □, Ni 1.00 at % Zr; ■, Ni 1.35 at % Cr; ◨, Ni 0.99 at % Ti.   (Ref. 29.)

lattice of the alumina to form a solid solution. In alumina-titanium composite systems, the Ti partially reduces the alumina fibers to form TiO, Ti$_3$Al, and an α-solid solution on the interfaces in sequence from the alumina side to the matrix side.

9.6 METALLIC-FIBER-REINFORCED METAL COMPOSITES

Metallic fibers or filaments for use in FRM include steel, stainless steel, molybdenum, and tungsten. Although these metallic fibers have relatively high densities compared with the nonmetallic fibers discussed earlier, they have such characteristics as high levels of wettability with matrix metals and high toughness. However, in most cases, the fibers have high reactivity and solubility values with matrix metals. From the viewpoint of the production of light-weight composite materials, reinforcement of light metals with metallic fibers is less advantageous than reinforcement with nonmetallic

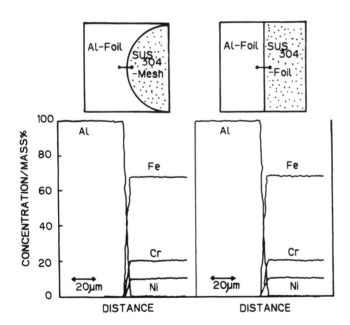

Figure 9.10   Concentration distribution across stainless steel wire or stainless steel foil in stainless steel/aluminum composites. (Ref. 31.)

fibers. However, when composite materials with high toughness and
strength levels are needed, this type of FRM is most dependable.

For the production of FRM, reinforcement of aluminum with steel
and stainless steel filaments has been considered. It has been observed that multilaminar reaction products are formed in both cases
[30,31]. Needless to say, the formation and growth of a reaction-
product layer are functions of temperature and time. For practical
use of this type of FRM, it is necessary to fabricate the FRM in
such a short time that the reaction-product layer is insufficiently
formed, and also at such a low temperature that growth of the layer
is out of the question.

Accordingly, it is expected that both steel and stainless steel
filaments have composite effects on the reinforcement of light metal
matrixes. When stainless steel/Al systems are formed by an explo-
sion pressing adhesion method, no elements are diffused through the
interface, and no intermetallic compounds or alloy layers are formed,

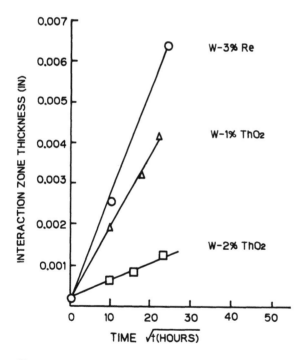

Figure 9.11 Rate of growth of filament-matrix interaction zone (at
2000°F) versus time for W−1% ThO_2/Mar M 322 (25% W), W−3%
Re/Mar M 322 (25% W), and W−2% ThO_2/Mar M 322 (25%W−E alloy)
composites. (Ref. 34.)

as shown in Figure 9.10 [31]. When an ultrahigh heat-resisting alloy is used as a matrix, tungsten fibers of high heat resistance and strength are effective as the reinforcing material. In most cases, tungsten is added with about 2% thoria (ThO_2) for inhibition of the embrittlement of the fibers caused by roughing and enlargement of the crystalline structure at high temperature. It has been reported that the addition of thoria has an effect on filaments of a simple tungsten substance, but in the case of FRM in which Ni- and Co-containing alloys are used as the matrix, Ni and the like are diffused into the tungsten filaments, resulting in embrittlement of the fibers.

The report states that CVD coating tungsten fibers with hafnium nitride (HfN) and hafnium carbide (HfC) as the intermediate layer is effective in solving such problems [32]. In fact, however, coating is not always effective for inhibition of the reactions. Hence it has been reported that the addition to tungsten filaments of 10% zirconium in place of thoria (ThO_2) is effective [33]. As stated above, when nickel-base alloys are used as the matrix, various interfacial reactions occur, but as shown in Figure 9.11, heat-resistant iron- and cobalt-base alloys have a less deleterious effect on the fibers than do nickel-base alloys [34].

REFERENCES

1. I. Shiota, Interface of Fiber Reinforced Metal, *Kogyo Zairyo*, *31*(13), 27 (1983).

2. P. G. Sullivan and L. Raymond, Graphite Fiber Morphology: Effect of Behavior in Graphite-Aluminium Composites, *10th Natl. SAMPE Tech. Conf.*, 466 (1978).

3a. M. F. Amateau, Progress in the Development of Graphite-Aluminium Composites Using Liquid Infiltration Technology, *J. Compos. Mater.*, *10*, 279 (1976).

3b. S. Kohara and N. Muto, Degradation Behavior of Carbon Fibers by Molten Aluminum, *J. Jpn. Inst. Met.*, *52*, 1063 (1988).

4a. L. W. Davis, Refractory Fiber Reinforcements for Elevated Temperature Applications, *Adv. Fiber Compos. Elevated Temp.*, 10 (1980).

4b. S. Kohara and N. Muto, Degradation Behavior of Carbon Fibers by Molten Aluminum, *J. Jpn. Inst. Met.*, *52*, 1063 (1988).

5a. S. Kohara, N. Muto, and Y. Imanishi, TiC Coating on Carbon Fiber and the Compatibility of TiC Coated Carbon Fiber with Aluminium, *J. Jpn. Inst. Met.*, *43*, 589 (1979).

5b. S. Paprocki, et al., Ultrahigh-Modulus Fibers: A New Dimen-
     sion for Graphite-Aluminum and Graphite-Magnesium Composites,
     24th Natl. SAMPE Symp. Exhib., 1451 (1979).

6.   A. Miyase and K. Piekarski, Compatibility of Chromium Car-
     bide Coated Graphites Fibers with Metalic Matrices, *J. Mater.
     Sci.*, *16*, 251 (1981).

7.   D. M. Goddard, Interface Reactions During Preparation of
     Aluminium-Matrix Composites by the Sodium Process, *J. Mater.
     Sci.*, *13*, 1841 (1978).

8.   J. Inagaki, Y. Terasawa, E. Nakata, and A. Okura, Mechanical
     Properties of Carbon Fiber Reinforced Aluminium Composites,
     Compos. Mater. II, 37 (1980).

9.   A. Kitahara, Surface Modification of Carbon Fiber Reinforced
     Aluminium Composites, *Kogyo Zairyo*, *33*(12), 66 (1985).

10. A. G. Kulkarni, B. C. Pai, and N. Balasubramanian, The
     Cementation Technique for Coating Carbon Fibers, *J. Mater.
     Sci.*, *14*, 592 (1979).

11. R. Warren, C. H. Anderson, and M. Carlsson, High-Tempera-
     ture Compatibility of Carbon Fibers with Nickel, *J. Mater. Sci.*,
     13, 178 (1978).

12. G. L. Ranshofen, Giesseigenshaften und Oberflächenspannung
     von Aluminium und binären Aluminiumlegierungen, *Aluminium*,
     49, 231 (1973).

13. Y. Kimura et al., Compatibility Between Carbon Fiber and
     Binary Aluminium Alloys, *J. Mater. Sci.*, *19*, 3107 (1984).

14. L. S. Gyzei et al., Interaction Between Fiber and Matrix of
     Carbon Fiber Reinforced Aluminium Composites, *Fiz. Khim.
     Obrab. Mater.*, *2*, 132 (1980).

15. I. Shiota and O. Watanabe, Compatibility Studies of Nickel
     Coated Carbon Fibers, *J. Jpn. Inst. Met.*, *38*, 794 (1974).

16. I. Shiota and O. Watanabe, High-Temperature Compatibility of
     Pyrolytic Carbon with Nickel-Copper Alloys, *J. Mater. Sci.,
     Lett. 14*, 1518 (1979).

17. Y. Kasai, M. Saito, and C. Asada, Application of New Sub-
     strateless Continuous SiC Filament to Metal Matrix Composite
     Materials, *J. Jpn. Soc. Compos. Mater.*, *5*, 56 (1979).

18. Y. Kasai and M. Saito, Application of New Substrateless Con-
     tinuous SiC Filament to Metal Matrix Composite Materials, *Compos.
     Mater. II*, 45 (1980).

19. S. Y. Kim, C. H. Lee, J. Yamamoto, and S. Umekawa, Interfacial Reaction Between SiC Fiber and Aluminium and the Effect of ZrC Barrier, *J. Jpn. Soc. Compos. Mater.*, 9, 22 (1983).

20. J. Cornie, R. J. Suplinskas, and A. Hauze, Filament Surface Enhancement for Casting and Hot Molding SiC—Al Composites, *Adv. Fibers Compos. Elevated Temp.*, 41 (1980).

21. N. N. Rykalin and M. Kh. Shorshorov, Physico-chemical Compatibility and Choice of Technology and Conditions for Producing Fiber Composite Materials, *Compos. Mater.*, 161, (1979).

22. P. Martineau, R. Pailler, M. Lahaye, and R. Naslain, SiC Filament/Titanium Matrix Composites Regarded as Model Composites, *J. Mater. Sci.*, 19, 2749 (1984).

23. E. M. Sokolovskaya, L. S. Gyzei, and L. L. Meshkov, The Rate of Interaction Between Titanium and SiC Fibers as Affected by Alloying, *Compos. Mater. II*, 62 (1980).

24. R. L. Mehan and R. B. Bolon, Interaction Between Silicon Carbide and a Nickel Based Superalloy at Elevated Temperatures, *J. Mater. Sci.*, 14, 2471 (1979).

25. E. M. Sokolovskaya, L. S. Gyzei, and L. L. Meshkov, On the Possibility of Reinforcement of Nickel Alloys by Non-metallic Fibers, *Compos. Mater. II*, 126 (1980).

26. A. Kelly and S. T. Mileiko, *Handbook of Composites*, Vol 4, *Fabrication of Composites*, Elsevier Science Publishing Co., Inc., New York, 1983, p. 221.

27. R. Suplinskas, J. Cornie, and A. Hauze, Boron Carbide Coated Boron for Metal Matrix (Al, Mg, Ti) Composites, *11th Natl. SAMPE Tech. Conf.*, 13, 1036 (Nov. 1979).

28. A. K. Dhingra, Metal Matrix Composites Reinforced with Fiber FP (α—Al$_2$O$_3$), *Philos. Trans. R. Soc. London*, A-294, 559 (1980).

29. W. H. Sutton, Wetting and Adherence of Ni/Ni Alloys to Sapphire, General Electric Co., *Report R-67*, SD 44, (1964).

30. S. Mito, Y. Ogino, and T. Okada, Constituent Distribution Around the Matrix-Fiber Interface of Stainless Steel Fiber Reinforced Aluminium Alloy Composites, *J. Jpn. Soc. Compos. Mater.*, 5, 31 (1979).

31. A. Chiba, M. Fujita, H. Tonda, and K. Hokamoto, Fabrication and Some Mechanical Properties of Explosively Compacted Stainless Steel/Aluminium Composites, *J. Jpn. Soc. Compos. Mater.*, 9, 108 (1983).

32. P. J. Mazzei, G. Vandrunen, and M. J. Hakim, Powder Fabrication of Fiber-Reinforced Superalloy Turbine Blades, *AGARD Conf. Proc.*, *200*, SC7-1 (1976).

33. G. Wirth, Comment on the Short Contribution No. 7, *AGARD Conf. Proc.*, *200*, C-1 (1976).

34. I. Ahmad and J. Barranco, $W-2\%$ ThO_2 Filament Reinforced Cobalt Base Alloy Composites for High Temperature Application, *Adv. Fibers Compos. Elevated Temp.*, 183 (1980).

10

Interfacial Effect of Carbon-Fiber-Reinforced Composite Material

10.1 INTRODUCTION

The linear complex effect, nonlinear secondary complex effect, the dispersion effect, interfacial effect, and so on, of fiber-reinforced composite materials are examined depending on modification of the reinforcing fibers and interface, and on the type of sizing agent and matrix resin used. Recently, interfaces between fibers and resins have been investigated from the viewpoint of a thickness rather than as a plane generally having no thickness. However, most of the "contents of interfaces" and bonding mechanisms have not been elucidated, because interfaces are ready to be stained and have extremely small thicknesses, if any. The roles or functions of the interfaces are different depending not only on the surface charac-teristics of the constituent materials but also on the characteristics of the materials in the bulk state. It is very difficult to describe generally the roles of interfaces. In this chapter we discuss the interfacial effect and the roles that interfaces or interlayers of car-bon-fiber-reinforced composite materials play in the strength of uni-directionally reinforced materails and laminates.

10.2 SURFACE TREATMENT OF CARBON FIBER

Carbon-fiber-reinforced plastics (CFRPs) can be given high levels of mechanical properties if they have sufficiently high bonding strength between the carbon fibers and matrix resins to transmit an applied stress.

In general, carbonization and graphitization (1000 to 3000°C) of carbon fibers are carried out in an inert ambience of nitrogen,

argon, and the like. The chemical composition and microstructure of
carbonized or graphitized fibers are changed depending on the firing
temperature. With increased firing temperature, the graphite crystalline sheath of the fiber grows larger and the orientation regularity
is enhanced [1].

The graphite crystalline sheathes are arranged over the surface
of each PAN carbon fiber. The surface layer of the graphite has a
tight structure with high crystallinity compared with the internal
layer of the graphite. The graphite surface is inactive and exhibits
low adhesive properties to epoxy resins and similar compounds. The
carbon fibers must be surface-treated for bonding, as described in
Chapter 9.

Figure 10.1 shows the results of ESCA of the carbon fibers as
to how the chemically active carbon fiber groups on the surface were
changed by oxidation treatment [1]. In the figure, the digital difference spectra clearly illustrate the changes in active groups. The
oxidation treatment of the carbon fiber caused functional groups such
as —COOH or —OH groups to be introduced onto the surface layer
of the carbon fiber [2].

Simultaneously, the tight surface layer of the carbon fiber was
partially broken, so that the more active edge portions of the graphite crystalline sheath of the fiber were exposed. Further, the surface area was increased, which improved the surface activity.

Composite materials reinforced with PAN carbon fibers oxidiza-
tion-treated with nitric acid were measured for interlaminar shear
strength (ILSS) with results shown in Figure 10.2 [3]. Oxidation
treatment increased by about 25% the ILSS of composite materials reinforced with carbonized fibers, and that of composite materials reinforced with graphitized carbon fibers to three times the ILSS of
composite materials reinforced with untreated graphitized carbon
fibers.

As described above, the oxidation treatment of carbon fiber is
effective for enhancing the bonding strength between carbon fiber
and resins. In an interesting experiment, the effect of the functional groups present on carbon fiber surfaces on the adhesive properties of fibers, particularly the ILSS in composite materails reinforced
with carbon fiber, was examined by removing functional groups from
the fiber surfaces using chemicals.

As is clear from Figure 10.3, in the case of methanol-HCl and
diazomethane used as chemicals, the methanol selectively blocked
the carboxyl groups, and the diazomethane blocked all the acidic
groups (carboxyl and phenol groups) on the fiber surface. Long-
time treatment with chemicals blocked all other functional groups.
Figure 10.4 illustrates the effects of chemicals in removing functional groups. Oxidation treatment of the carbon fiber increased the
number of acidic functional groups, resulting in increases in bonding

C₁s difference spectra

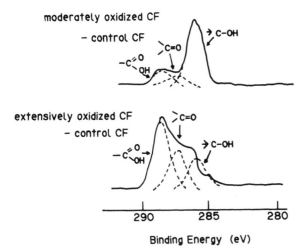

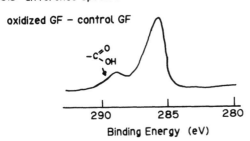

Figure 10.1 Digital difference spectra of surface-oxidized carbon fiber (CF) and graphite fiber (GF). (Ref. 1.)

Interlaminar shear strength (X 10³ psi)
14
12
10
8
6
4
2
0
10
20
30
40
50
60
70
Time (hours)
Type II(S)
Type I (S)
L/d=6

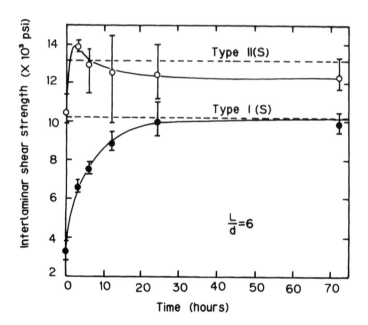

Figure 10.2 Effect of nitric acid oxidation time on the interlaminar strength of CFRP reinforced with treated Modmor fibers (Type I graphitized PAN based; Type II, carbonized PAN based). (Ref. 3.)

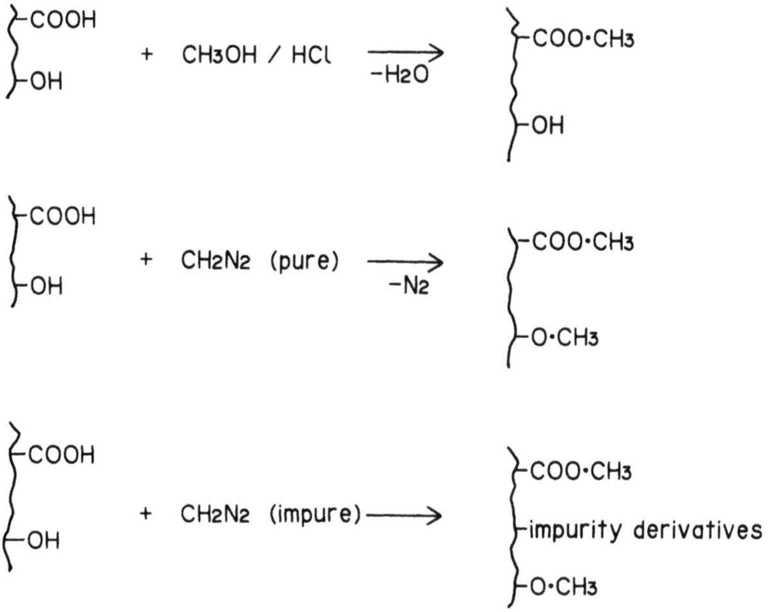

COOH
OH
+ CH3OH / HCl
-H2O
COO·CH3
OH
COOH
OH
+ CH2N2 (pure)
-N2
COO·CH3
O·CH3
COOH
OH
+ CH2N2 (impure)
COO·CH3
impurity derivatives
O·CH3

Figure 10.3 Elimination of functional groups on the carbon fiber surface by the reagent. (Ref. 4.)

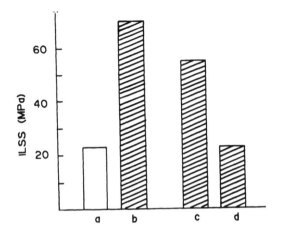

Figure 10.4   Relation between elimination of functional groups on the carbon fiber surface by diazomethane and interlaminar shear strength. a, graphite fiber; none treated; b, graphite fiber; HNO_3 treated at 120 h; c, pure diazomethane-treated oxidized graphite fiber; d, impure diazomethane-treated oxidized graphite fiber. (Ref. 4.)

strength between the carbon fiber and the resins. In turn, when the resultant functional groups were blocked by methylation treatment, the bonding strength between the carbon fiber and the resins was remarkably reduced, and approached that between untreated carbon fiber and resins.

The test results above suggest that the bonding strength between carbon fibers and resins is dominated by chemical bonds between the resin and the functional groups on the carbon fiber surface. As functional groups present on the carbon fiber surface, carboxyl and acidic hydroxyl groups, quinones, and so on could be named. It is suggested that the functional groups are bonded to the resins in the form of ether bonds, hydrogen bonds, and so on [4,5], as shown in Figure 10.5.

For surface treatment of the carbon fiber, a vapor coating method is effective, where the carbon fiber is vapor-coated with carbon, SiC, polymers, and so on. Acetylene was decomposed at 1100 to 1200°C, so the carbon fiber was coated with the resultant carbon. As shown in Figure 10.6, the ILSS of the CFRP was increased when the carbon coating amount was 0 to 25% [6].

According to one surface treatment method, whiskers of SiC and the like were formed on carbon fiber surfaces. The CFRP reinforced with the carbon fiber coated with whiskers had a high ILSS value [7]. In another method, carbon fiber was etched by hydrogen or oxygen

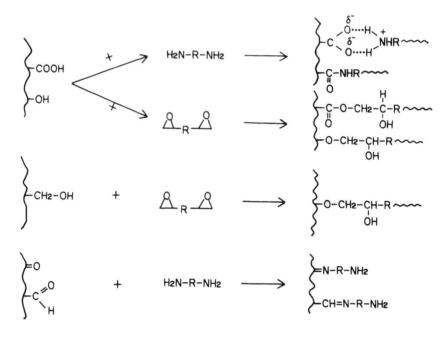

Figure 10.5   Reaction between functional groups on the carbon fiber surface and epoxy matrix or amine catalyst.   (Ref. 4.)

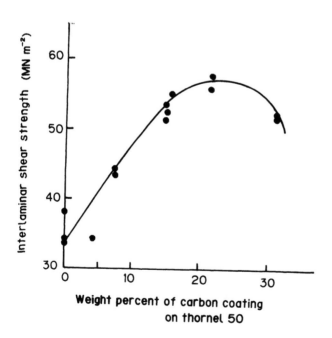

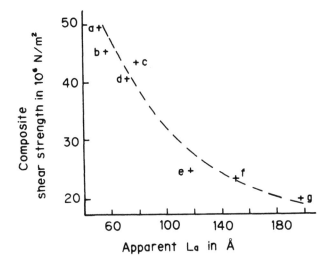

Figure 10.7 Relation between composite shear strength and the L_a found through the Raman spectrum of the graphite fiber. A, Morganite II (untreated); b, Thornel 25 nitric acid oxidized; c, Hitco H.M.G. 50 (untreated); d, Thornel 25 air oxidized (400°C); e, Thornel 25 (untreated); f, Thornel 40 (untreated); g, Morganite I (untreated). (Ref. 9.)

plasma discharging, followed by plasma discharge treatment of the carbon fiber in a monomer vapor ambience, so that the carbon fiber was coated with polymers of acrylonitrile, MMA, and so on [8a]. For the purpose of enhancing the reinforcement-matrix interaction in carbon fiber reinforced polymer composite, mechanical and spectroscopic studies were made on the epoxy resin composite reinforced with the carbon fiber coated with thin layer of polyimide [8b]. The peel strength of a laminated specimen and the fiber efficiency factors for modulus and strength are larger than those of the composite reinforced with nonpolyimide treated fiber. The occurence of specific interaction between an epoxy resin and the polyimide were recognized on FTIR.

Figure 10.6 The interlaminar shear strength shows a maximum around 20 to 25 wt % of deposited carbon. These results were obtained in 20 vol % acetylene. Slightly higher shear strengths were obtained with 80 vol % acetylene. No runs were conducted in pure acetylene or with other types of fibers. The volume fraction of the Thornel 50 yarn was held constant at 0.5. (Ref. 6.)

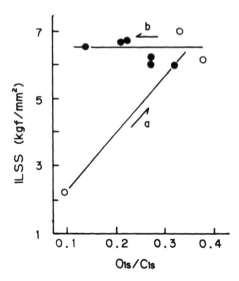

Figure 10.8   Relation between the concentration of oxygen-containing groups in Torayca M40A carbon fiber and the ILSS of its CFRP.   a, surface treatment; b, heat treatment in vacuum.   (Ref. 1.)

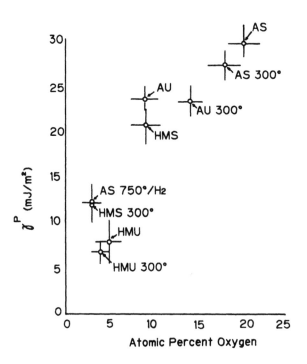

Figure 10.9   Plot of the polar component of graphite fiber surface free energy (γ^P) versus XPS-determined oxygen content for all graphite fibers and treatments studied.   (Ref. 11.)

There are some experimental facts which suggest that the bonding strength between the carbon fiber and resins does not depend on chemical bonds alone. For example, when the area extending from the graphite laminar plane surrounding a surface of the carbon fiber is smaller, the ILSS of the CFRP tends to be higher. The test results shown in Figure 10.7 suggest that the bonding strength between the carbon fiber and resins is increased when the number of carbon atoms present on the periphery of one graphite plane (i.e., the number of active carbon atoms contained in the peripheral carbon atoms) is increased [9].

The amount of oxygen present on the carbon fiber surface can be measured by ESCA [10]. Figure 10.8 shows a plot of the ILSS of the CFRP reinforced with PAN graphitized fibers against the ESCA oxygen concentration of carbon fiber surfaces [1]. The ILSS was increased with an increase in oxygen concentration caused by oxidation treatment of the carbon fiber.

In the case of carbon fiber heated under vacuum, although the oxygen concentration was reduced, the ILSS was substantially unchanged. This experimental result implies that the mechanical bonding between the carbon fiber and resins (i.e., engagement of

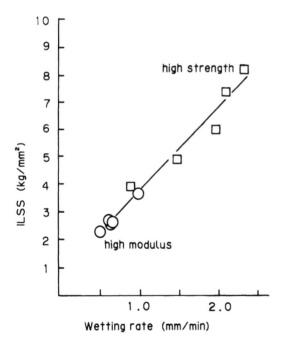

Figure 10.10   Relation of interlaminar shear strength to wetting rate of liquid resin into carbon fibers.   (Ref. 12.)

interlock of the resins with the concave surfaces of the carbon fibers produced by the oxidation) contributes more to the bonding between them than does chemical bonding between the resins and oxygen-containing functional groups present on carbon fiber surfaces.

As described in Chapter 2, the wettability of carbon fiber surfaces with liquid resins is an important factor concerning bonding between carbon fiber and resins. In some investigations, the contact angles to carbon fiber surfaces of a series of liquids of which the total free energy, dispersion, and polar constituents are known were measured to evaluate independently the surface free energy, dispersion constituents, and polar constituents of carbon fiber [11]. Figure 10.9 shows the test results. With an increase in oxygen content on the carbon fiber surfaces, the total free energy of the carbon fiber surfaces was increased. The increase in total free energy is attributed to the increase in number of polar constituents, not to the number of dispersion constituents. In fact, the ILSS of CFRP reinforced with carbon fiber having higher wettability was enhanced, as shown in Figure 10.10 [12].

10.3 ROLES OF INTERFACE IN MECHANICAL PROPERTIES OF COMPOSITES

In most cases the mechanical properties of fiber-reinforced composite materials are analyzed and estimated on the basis of the characteristics of fundamental unidirectionally reinforced materails. In the model for unidirectionally reinforced composite materails shown in Figure 10.11, the interfaces inherent in such materials exert a strong influence on F_T (the tensile strength of the material perpendicular to the fiber axis) and $F_{LT, \phi}$ (the shear strength of the material along the common plane of the fibers). That is, the interfacial strength between the fibers and resins, as well as the voids generated during molding, have a direct influence on the strength of materials in the T direction (F_T).

On the other hand, the matrix resin of a composite material is hindered from even curing or shrinking by the fibers present in the composite material. Generally, the resin itself has anisotropic properties in the composite material, and the mechanical properties of the resin are different from those of the resin in the bulk state. Also, in the case of high-strength fibers to reinforcing resins, sufficient bonding between the matrix resins and the fibers at the interfaces in indispensable for realizing the potential of the matrix resins.

If there are deficiencies in bonding between the resins and the fibers or in the impregnation of the resins into the fibers, stress will be concentrated on the fibers, which leads to breaking of the entire composite. In fact, as shown in Figure 10.12, F_T and the F_{LT} are about 1/30 to 1/50 of F_L' (the tensile strength in the fiber

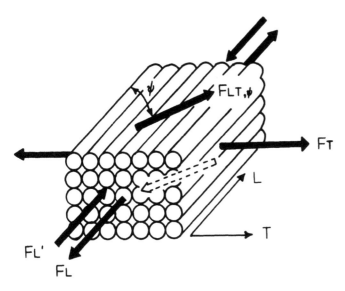

Figure 10.11 Schematic representation of unidirectional fiber-reinforced plastics. (Ref. 13, p. 228.)

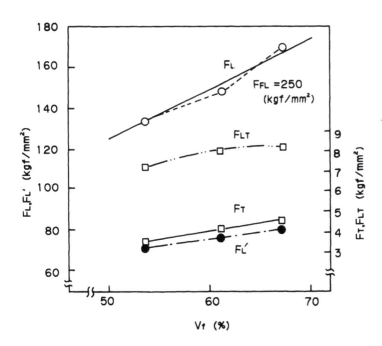

Figure 10.12 Relation between various strength of unidirectinal
carbon fiber-reinforced plastics and V_f. (Ref. 13, p. 228.)

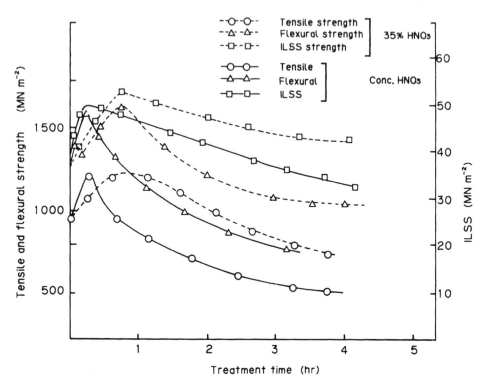

Figure 10.13 Variation of tensile strength, flexural strength, and interlaminar shear strength of composites with HNO₃ treatment time of carbon fibers LY556 + HY951 system and LY556 + HT972 system. (Ref. 15.)

axial direction) and F'$_L$ (the compression strength in the fiber axial direction). Generally, breaking at the initial and maximum loads of such composite materails is attributed to F$_T$ and F$_{LT}$, not to F$_L$ [13,14].

Accordingly, the reinforcing fibers are surface treated, correspondingly to each type of fiber, as described in Chapter 9 for enhancement of the strength of composite materials in the fiber axial direction. Figure 10.13 illustrates changes in tensile strength, flexural strength, and interlaminar shear strength of composite materials whose reinforcing carbon fibers have been treated with nitric acid [15]. As shown in the figure, treatment with 35% nitric acid improved the strength of the composite materials, which is evidence of the effectiveness of nitric acid oxidation treatment of carbon fibers.

Although the number of active groups present on fiber surfaces increases with intensified oxidation treatment, improvement in the characteristics of composite materials is limited to some degree. It is supposed that further improvement in the characteristics of materials is dominated by the properties of the matrix resin rather than the interfacial characteristics. However, it has not been determined quantitatively whether or not the reactions between functional groups present on fiber surfaces and matrix resins are saturated. This and similar matters are subjects for further study.

Many attempts have been made to improve the adhesive properties of carbon fibers to matrix resins using coatings of oxidized carbon fibers with various polymers. It has been reported that composite materials reinforced with such polymer-coated carbon fibers showed improved interlaminer shear strength. Table 10.1 shows an example [16].

Consideration needs to be given to interfaces between carbon fibers and matrix resins from the viewpoint of even distribution of the fibers in the matrix resins, in addition to the bonding between them. More particularly, in a composite material composed of fibers and a matrix resin, the fibers are unevenly distributed in the matrix resin. The fibers in the composite material are situated close to each other or in contact locally. It is supposed that stress concentration occurs on such portions, where the fibers are close or in contact. To reduce this stress concentration, fibers coated with polymers and matrix resins were combined so as to prevent the fibers from coming into contact, resulting in a significant improvement in the physical properties of the composite material [17].

Table 10.1   Effect of Polymer Coating on Shear Strength of Composite

Oxidation	Polymer coating	Polymer (%)	Density (g/cm^3)	Shear strength (psi)
None	None	—	1.28	2350
60% HNO_3, 24 h	None	—	1.29	3520
	PVA	7	1.31	6200
	PVC	7	1.31	6100
	Rigid poly-urethane	3	1.27	5900
	PAN	7	1.27	2400

Source: Ref. 16.

Table 10.2 Properties of Whiskerized Graphite Laminates in Epoxy Matrices

Matrix	Fiber	V_f (%)	Void (%)	Whisker-izing (vol %)	Flexural strength ($\times 10^3$ psi)	Flexural modulus ($\times 10^6$ psi)	Shear strength ($\times 10^3$ psi)
Epon 828/1031/ NMA/BDMA	Morganite I	60	0.5	5.5	84.9	33.1	13.9
		59	0	1.6	89.9	32.1	6.9
		62	3.7	6.9	77.9	16.4	Tensile break at (6.5)
	Thornel 50	60	9.0	7.4	40.5	16.6	Tensile break at (8.5)
		62	0.8	3.9	83.8	15.9	Tensile break at (7.8)
DEN 438/ NMA/BDMA	Morganite I	59	0	4.6	85.0	32.3	10.8
		63	0.8	4.9	85.2	16.3	(Tensile break at (8.2)

Source: Ref. 20.

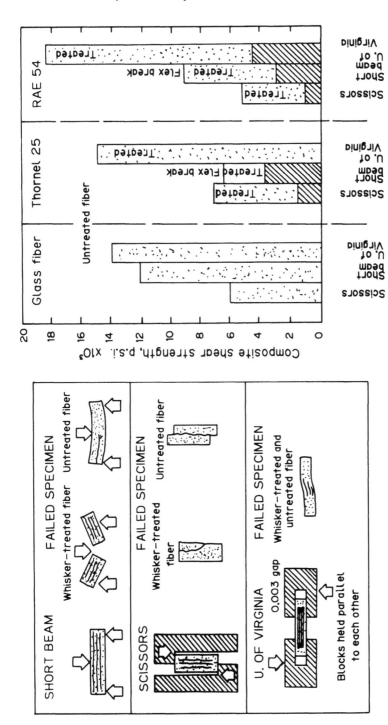

Figure 10.14 Shear tests, showing type of failure and summary of shear test results. (Ref. 19.)

In addition to the surface treatment methods for carbon fibers discussed in Chapter 8, hydrolysis carbon coating, whiskerizing, and other methods are known. The mechanical properties of composite materials utilizing such treated fibers have been examined [18].

For example, silicon carbide whiskers were deposited on carbon fibers in the range 0.5 to 3 wt %. As seen in Figure 10.14 and Table 10.2, the shear strength of composite materials reinforced with whiskerized carbon fibers was improved [19,20]. However, the test results in Figure 10.15 reveal that the shear strength of composite materials is largely influenced by voids generated during molding.

The shear strength of composite materials utilizing treated carbon fibers was improved significantly when single crystals of β-silicon carbide were grown on the surface of the carbon fibers perpendicular to the fiber axis. As shown in Figure 10.16, the shear strength of composite materails reinforced with treated Thornel 25 (a rayon-type carbon fiber) and with treated PAN carbon fibers was enhanced about threefold and fivefold, respectively, over the shear strength of composite materials reinforced with untreated carbon fibers [21a].

Recently, the carbon fiber reinforced epoxy resin modified with liquid carboxyl terminated nitrile rubber proved on effective method of interfacial adhesion [21b]. The increase of peel strength by adding liquid carboxyl terminated nitrile rubber are suggested to depend on

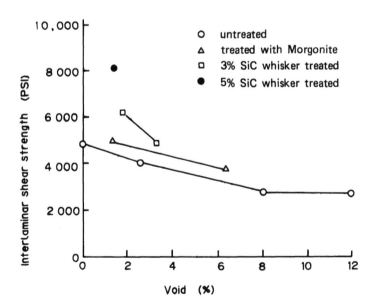

Figure 10.15 Effect of void content and surface treatment on shear strength of polyimide-Morgonite Composites. (Ref. 20.)

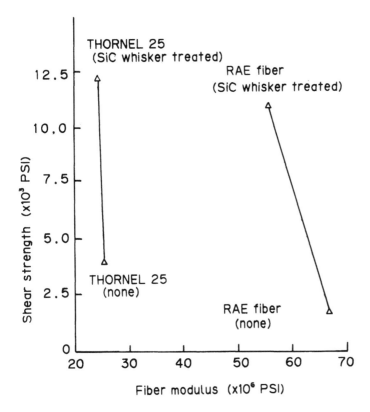

Figure 10.16   Effect of SiC whisker treatment on shear strength of composite. (Ref. 21a.)

the force used to pull out these rubber microparticles from a scanning electron microscopic analysis.

10.4   ROLES OF INTERFACE IN PHYSICAL PROPERTIES OF LAMINATED CFRP

When a light weight structural material is needed, CFRP, which has high specific strength and rigidity, is used. However, CFRP is disadvantageous in that it has a substantially linear correlation between stress and strain until F_L and F_T breaking occurs. In addition, the breaking strain is low, up to 1%. Unlike metal materials, CFRP, exhibits no plasticity and ductility. Accordingly, to achieve the characteristics of the materials, unidirectionally carbon-fiber-reinforced composite materials are laminated in several directions for use.

Needless to say, sufficient interlaminar bonding strength and toughness are required for laminates. For example, when the bonding strength and toughness between layers of a laminate are insufficient, the ends of the laminate are delaminated prior to breaking of the entire laminate, even by loading in the center of a plane of the laminate, since each layer of the unidirectionally reinforced materail has highly nonisotropic properties, which even causes such loading to generate delamination or peeling stress between the layers at the ends. Thus, in this case, the potential characteristics of a laminate of unidirectionally reinforced materails cannot be sufficiently well realized for use.

Various studies have been made on the optimization of laminate construction [22]. One serious problem is that considerable damage is generated when an impact load of low energy is applied to the sides and corners of a laminate, reduceing its compression strength [23].

To solve such a problem, it is necessary to improve the interlayer toughness.

1. If high bonding strength between the fibers and the matrix resins at the interfaces can be obtained, it is first necessary to enhance the toughness of the matrix resins to form a layer of laminates. The following methods are effective for improvement:

   a. Development of novel thermosetting resins that have a level of toughness [24]
   b. Modification of epoxy resin by blending together thermoplastic resins [25]
   c. Using as the matrix resin thermplastic resins that have a high level of elongation, such as polyether sulfone (PES) and polyether ether ketone (PEEK) [26]

Applied in practice, these methods caused significant improvement in the compression strength of the laminate after it was subjected to impact, as shown in Figure 10.17. These experimental results are but one example showing that the residual compression strength following impact testing was improved by enhancing the toughness of matrix resins to reduce the interlayer damage caused by impact. Generally, thermoplastic resins having high elongation exhibit high strength values, irrespectively of T_g, crystallinity, and amorphous properties of the resins [27a]. In particular, it was reported that polyether ether ketone (PEEK) has the highest residual compression strength after impact testing of thermoplastic resins. Further, adhesion between carbon fiber and thermoplastic matrices with heat distortion temperature is attributed to the differential thermal shrinkage of fiber and matrices [27b].

2. The interlayer toughness of composites is improved by using a hybrid laminate constitution. Typically, ductile fibers such as glass and aramid fibers are used together with carbon fibers to improve the breaking toughness of laminates. For the viewpoint of laminate materials, a CFRP layer is laminated to a glass-fiber or

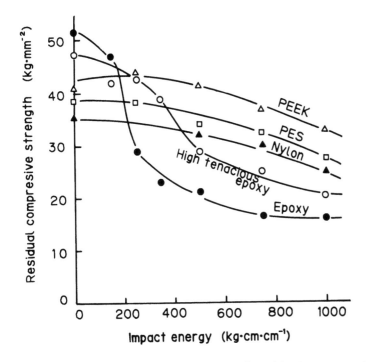

Figure 10.17   Matrix dependence of residual compressive strength after impact test.   (Ref. 27a.)

aramid-fiber-reinforced material layer.   Otherwise, CF/GF mixed fibers are used as a reinforcing material, and so on.

Figure 10.18 represents the stress-strain curve of a FG/CF unidirectionally, reinforced composite material.   The breaking strain ε_c of CFRP is smaller than that of GFRP.   If there is no bonding effect between the CFRP and the GFRP, the CFRP will be broken at the point of breaking strain (σ_c is the initial breaking stress) according to the mixing law as shown by the solid line in the figure, and after a reduction in the load, the GFRP will only withstand the load to the ε_g point, which is followed by breaking of the reinforced material.

However, if both fibers bond sufficiently well to the matrix resin, when the carbon fiber begins to be partially broken, the glass fibers situated near the breaking portions share the load through shearing of the matrix resin, and the cut carbon fibers are effective as reinforcing short fibers.

Measurement of the initial breaking stress σ_{CG} at a ε_{CG} value higher than ε_C are shown as black circles in Figure 10.18.   Thereafter, when the carbon fibers are sequentially broken, the fibers

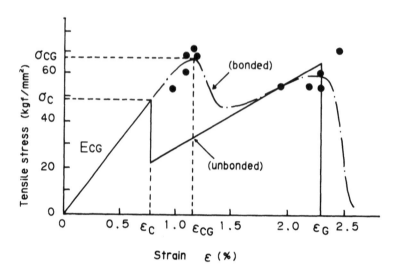

Figure 10.18 Stress-strain curves of unidirectional fiber-reinforced hybrid composite. (Ref. 13, p. 220.)

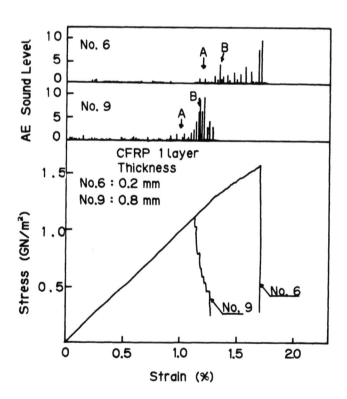

Figure 10.19 Stress-strain curve and AE sound level versus strain curve in carbon and aramid hybrid composites. (Ref. 29.)

Table 10.3 Impact Properties of Unidirectional Composite Materials as Determined from Instrumented Charpy Test

Reinforcing fibers	Apparent flexure strength [ksi (MN/m^2)]	Total energy per unit area [ft·lb/in^2(J/m^2)]		Ductility index
		Dial	Oscilloscope	
E glass	73 (500)	114 (2.4 × 10^5)	114 (2.4 × 10^5)	0.4
Kevlar 49	142 (980)	124 (2.6 × 10^5)	114 (2.4 × 10^5)	24 (1.6)[a]
HMS-Graphite	125 (860)	3.8 (8 × 10^3)	3.8 (8 × 10^3)	0.0
20% Kevlar 49 80% HMS-Graphite	170 (1170)	34.3 (7.2 × 10^4)	30.5 (6.4 × 10^4)	6
41% Kevlar 49, 59% HMS-Graphite	141 (970)	46.7 (9.8 × 10^4)	42.9 (9 × 10^4)	4

[a]The first value was based on the onset of nonlinearity. The numbers in parentheses are based on maximum stress.
Source: Ref. 30.

are elongated with a reduction in the load. Conclusively, the break-ing toughness of the reinforced material is enhanced [28].

When CFRP layers and aramid-fiber-reinforced layers (AFRP) are hybrid-laminated together, it is preferred to make the layers thinner and to laminate a larger number of layers, since the breaking stress and breaking strain are increased, which leads to improvements in the breaking toughness of the laminate. Figure 10.19 shows the test results [29].

This fact is attributed to the AFRP layers, which prohibit the propagation of cracks in the CFRP layers. Table 10.3 tabulates comparisons of the Charpy impact energy with the ductility index (DI) of the laminates, indicating that the hybrid structure laminates have a DI about three times as high as that of laminates reinforced with single aramid fibers [30].

However, in the case of multidirectionally laminated boards, stress is concentrated on the interlayers. Test results of bending strength have revealed that multidirectionally laminated boards are deficient in interlaminar shear strength [31]. As reported previously, the hybrid laminate deficiency—readiness to be delaminated—is attributed to differences in the modulus of elasticity and heat expansion ratio of the fibers used as hybrids, and also to differences in residual shrink-age after lamination, which generate heat residual stress and three-dimensional stress in the laminates [32]. For the production of hy-brid laminate boards, consideration must be given to the surface treatment of different types of fibers to improve their interfacial ad-hesive properties to matrix resins and also to reducing the internal stress generated.

REFERENCES

1. K. Morita, Y. Murata, A. Ishitani, K. Murayama, T. Ono, and A. Nakajima, Characterization of Commercially Available PAN Based Carbon Fibers, *Pure Appl. Chem.*, *58*, 455 (1986).

2a. A. Proctor and P. M. A. Sherwood, X-Ray Photoelectron Spec-troscopic Studies of Carbon Fiber Surfaces: III. Industrially Treated Fibers and the Effect of Heat and Exposure to Oxygen, *Surf. Interface Anal.*, *4*, 212 (1981).

2b. A. Proctor and P. M. A. Sherwood, X-Ray Photoelectron Spec-troscopic Studies of Carbon Fiber Surfaces: II. The Effect of Electrochemical Treatment, *Carbon*, *21*, 53 (1983).

2c. A. Proctor and P. M. A. Sherwood, Data Analysis Technics in X-Ray Photoelectron Spectroscopy, *Anal. Chem.*, *54*, 13 (1982).

3. D. W. McKee and V. J. Mimeault, Surface Properties of Carbon Fibers, *Chem. Phys. Carbon, 8*, 151 (1973).

4a. E. Fitzer et al., *Ext. Abs. 16th Bien. Conf. Carbon*, p. 494 (1983).

4b. K. Kubo, M. Koishi, and T. Tounoda, Ed., *Composite Materials and Interface*, Sogo Gijutsu Shuppan, Tokyo, 1986, p. 257, 258 (A. Shinds).

5. K. Horie, H. Murai, and I. Mita, Bonding of Epoxy Resin to Graphite Fibers, *Fibre Sci. Technol.*, 9, 253 (1976).

6. D. J. Pinchin and R. T. Woodhams, Pyrolytic Surface Treatment of Graphite Fibers, *J. Mater. Sci.*, 9, 300 (1974).

7. M. G. Busche, High Modulus Carbon Fiber: New Entry in FRP Race, *Mater. Eng.* (Feb.), 46 (1968).

8a. A. Benatar and T. G. Gutowski, Effects of Surface Modification of Graphite Fibers by Plasma Treatment on the Hygral Behavior of Composites, *39th Annu. Tech. Conf. SPI*, 3-F (1984).

8b. M. Kodama, I. Karino, and J. Kobayashi, Interaction between the reinforcement and matrix in carbon fiber reinforced composite; Effect of forming the thin layer of polyimide resin on carbon fiber by In Situ polymerization, *J. Appl. Polym. Sci.*, 33, 361 (1987).

9. F. Tuinstra and J. L. Koenig, Characterization of Graphite Fiber Surfaces with Raman Spectroscopy, *J. Compos. Mater.*, 4, 492 (1970).

10. A. Ishitani, Application of X-Ray Photoelectron Spectroscopy to Surface Analysis of Carbon Fiber, *Carbon*, 19, 269 (1981).

11. G. E. Hammer and L. T. Drzal, Graphite Fiber Surface Analysis by X-Ray Photoelectron Spectroscopy and Polar/Dispersive Free Energy Analysis, *Appl. Surf. Sci.*, 4, 340 (1980).

12. M. Yamamoto, Surface Treatment Methods of Carbon Fiber, *Ext. Abs. Int. Symp. Carbon*, 3A05, 292 (1982).

13. M. Uemura, H. Kausai, H. Maki, and O. Watanake, Ed., New Composite Materials and Advanced Technology, Tokyo Kagaku Dojin, 1986.

14a. M. Uemura and H. Iyama, Rigidity and Strength of Quasi-isotorpic CFRP Laminated by Finite Number of Laminas, *J. Soc. Mater. Sci. Jpn.*, 33, 91 (1984).

14b. M. Uemura, H. Iyama, and Y. Yamaguchi, Discussion on Thermal Residual Stress and Buckling of Filament-Wound Circular Disc, *Rep. Inst. Space Aeronaut. Sci. Univ. Tokyo*, 17(1B), 341 (1981).

15. L. M. Manocha, Role of Fiber Surface-Matrix Combination in Carbon Fiber Reinforced Epoxy Composites, *J. Mater. Sci.*, *17*, 3039 (1982).

16. J. W. Beard and S. P. Prosen, Unidirectional High Modulus Graphite Composites, *24th Annu. Tech. Conf. SPI*, 2-B (1969).

17. L. D. Tryson and J. L. Kardos, The Use of Ductile Inner-layers in Glass Fiber Reinforced Epoxies, *36th Annu. Tech. Conf. SPI*, 2-E (1981).

18a. S. Ochiai and Y. Murakami, Tensile Strength of Composites with Brittle Reaction Zones at Interfaces, *J. Mater. Sci.*, *14*, 831 (1979).

18b. S. Ochiai and Y. Murakami, Difference in Tensile Strength Among Rayon-, PANI- and PAN II-Type Graphite Fibers After Sodium or Titanium-Boron Treatment, *Metall. Trans.*, *12A*, 684 (1981).

19. R. A. Simon and S. P. Prosen, Properties of Carbon Fiber Composites: Effect of Coating with Silicon Carbide, *Mod. Plast.* (Sept.), 227 (1968).

20. R. G. Shaver, Laminates and Filament-Wound Structure of Whiskerized High-Modulus Carbon Fiber, *24th Annu. Tech. Conf. SPI*, 15-B (1969).

21a. B. McCarroll and D. W. McKee, Interaction of Atomic Hydrogen and Nitrogen with Graphite Surfaces, *Nature*, *225*, 722 (1970).

21b. K. Nakao and T. Yamaguchi, Effect of Modification with Liquid Carboxyl Terminated Nitrile Rubber on Peel Strength of Carbon Fiber/Tetrafunctional Epoxy Resin Composite, *Kobunshi Ronbunshu*, *45*, 1 (1988).

22a. H. Fukunaga, Stiffness Characteristics and Optimization of Orthotropic Laminates, *J. Jpn. Soc. Compos. Mater.*, *11*, 27 (1985).

22b. J. Onoda, Optimal Laminate Configurations of Cylindrical Shells for Axial Buckling, *AIAA J.*, *23*, 1093 (1985).

22c. I. Susuki, Ply Thickness Optimization of Symmetric Laminates Under In-Plane Loadings, *30th Natl. SAMPE Symp.*, 19 (Mar. 1985).

22d. K. Ikegami, Y. Nose, T. Yasunaga, and E. Shiratori, Failure Criterion of Angle Ply Laminates of Fiber Reinforced Plastics and Applications to Optimise the Strength, *Fibre Sci. Technol.*, *16*, 175 (1982).

23. M. D. Rhodes, J. G. Williams, and J. H. Starnes, Jr., Low-Velocity Impact Damage in Graphite-Fiber Reinforced Epoxy Laminates, *34th Annu. Tech. Conf. SPI, 20-D* (1979).

24. T. Tattersall, Definition of the General Requirements for Composite Performance Parameters in Primary Structures, *5th Eur. SAMPE (Mater. Processes 5th Technol.)* (1984), p. 15.

25. T. F. Tanes, New High Strain to Failure Structural Graphite Preprog for Primary Aircraft Structures, *5th Eur. SAMPE (Mater. Processes 5th Technol.)* (1984) p. 5.

26. D. R. Carlile and D. C. Leach, Damage and Notch Sensitivity of Graphite/PEEK Composite, *15th Natl. SAMPLE Tech. Conf.,* 82 (1983).

27a. M. Kitanaka, H. Kobayashi, T. Norita, and Y. Kawatsu, Damage Tolerance of Thermoplastics/Graphite Fiber Composites, I.C.C.M.−V, 913 (1985).

27b. L. Dl, Landro, and M. Pegoraro, Carbon Fiber-Thermoplastic Matrix Adhesion, *J. Mater. Sci.,* 22, 1980 (1987).

28. T. Hayashi, K. Koyama, A. Yamazaki and M. Kihara, Development of New Material Properties by Hybrid Composition, *Fukugo Zairyo,* 1, 21 (1972).

29. S. Amagi and Y. Miyano, Tensile Properties of Carbon/Aramid Hybrid Unidirectional Reinforced FRP, *J. Jpn. Soc. Compos. Mater.,* 11, 62 (1985).

30. P. W. R. Beaumont, P. G. Riewald, and C. Zweben, Methods for Improving the Impact Resistance of Composite Materials, ASTM Symp. Compos. Mater., *ASTM STP-568,* American Society for Testing and Materials, Philadelphia, 134, 1974.

31. M. Uemura, H. Iyama, and H. Fukunaga, *Proc. 1st Int. Conf. Compos. Struct.,* Paisley, 282 (1983).

32. H. Ota, G. Hen, and M. Uemura, Bending Strength and Buckling of Filament Wound Cylindrical Shell, 15th FRP Symp., *J. Soc. Mater. Sci. Jpn.,* March 18, Osaka (1986).

11

Interface Analyses of Composite Materials

11.1 INTRODUCTION

The properties of composite materials are determined by each property of the filler or matrix and also by the structure of the interface between the two. It is a fundamental feature of composite materials and one of the most important problems in the revelation of its functions that there is, between the different phases, a surface having a special structure and functions.

For fiber-reinforced metals, for instance, the low affinity between the matrix and the reinforcing fiber will result in a decreased interfacial reaction zone and will have some influence on its moldability. On the other hand, a high affinity will cause an increase in the interfacial reaction zone, which will then significantly affect the properties of the composite materials. The same phenomenon is observed in the case of fiber-reinforced plastic (FRP), and it is known that there is along the interface a special area, which we refer to simply as the "interface," which is different in physical and chemical structures from those of the raw materials themselves.

Along the solid-phase interface, substances of various chemical species will contact and combine; however, the state of the junction will differ in accordance with the type of bonds constituting the chemical species: that is, primary bonds (ionic bond, covalent bond, metallic bond) or secondary bonds (hydrogen bond, etc.). In other words, the type of interaction along the interface will exert a great influence on the various properties of the composite material. Therefore, to improve the performance of composite material, it is absolutely necessary to clarify the structure of the interface, what is occurring there, and the role to be played by the interface.

Table 11.1   Analytical Method of the Composite Interface by Various Analytical Apparatus[a]

Analytical method[b]	Surface treated agents (STA)	Surface of STA layer	STA/Matrix interface[c]	Filler/Matrix interface[c]
AES		◎	○	○
ESCA		◎	○	○
SIMS	○	○	○	○
ISS	○	○	○	○
Laser Raman	◎	◎	◎	◎
IR (ATR, FT-IR)	◎	◎	◎	◎
X-ray fluorescence		○	○	○
UV resonance Raman		○	○	○
EM, SEM, SAM		◎	◎	◎

[a] ○, application for model sample; ◎, application for the real sample.
[b] AES, Auger electron spectroscopy; ESCA, electron spectroscopy for chemical analysis; SIMS, secondary ion mass spectrometry; ISS, ion scattering spectroscopy; EM, electron microscopy; SEM, scanning electron microscopy; SAM, scanning auger microscopy.
[c] Application for fracture surface, also.
Source: Ref. 1.

Methods for analysis of both the interface and the surface of the composite material are shown in Table 11.1 [1]. A few application methods are described here as typical examples of instrumental analysis.

11.2 ANALYSES OF THE SURFACE AND THE INTERFACE OF MATRIX

The information obtained from an application of infrared spectroscopy to surface (interface) investigation includes the molecular structure, orientation, chemical reaction, conformation, crystallinity, and so on, of the surface; thus a large amount of information is obtained. The usual methods for observation by infrared absorption spectrum are as follows:

1. The abrasion method.
2. The ATR method (attenuated total internal reflection).
3. The MIR method (multiple internal reflection).

There are many research reports on surface analysis done using these methods [2–6]: however, as the measured depth is at most only about 2 μm, surface analysis is often felt to be insufficiently sensitive.

Figure 11.1 shows, for example, the ATR spectra of the etched polyethylene surface treated with a chromic acid group [6]. Absorption bands due to surface treatment appear at 3300, 1700, 1260, 1215, and 1050 cm^{-1}; 3300 cm^{-1} represents the absorption due to the hydroxyl group and 1700 cm^{-1} that due to the carbonyl group; and 1260, 1215, and 1050 cm^{-1} are all due to the alkyl sulfonate group

$$CH_3-O-\overset{\overset{O}{\|}}{\underset{\underset{O}{\|}}{S}}-O-CH_2-\overset{\overset{O}{\|}}{\underset{\underset{O}{\|}}{S}}-O^-$$

The formation of these polar groups contributes increased adhesion, as shown in Figure 11.2 [6].

ESCA is widely used at present in the surface analysis of polymer (matrix resin). The surface designated here is the surface layer through which photoelectrons discharged from C_{1S} or O_{1S} can escape to the outside from points near the surface; the analysis here includes the analysis of elements and the qualitative analysis of functional groups. ESCA can readily be used to analyze

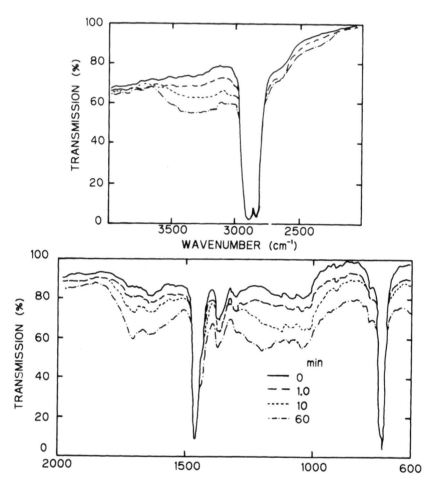

Figure 11.1 Surface IR spectra of etched LDPE-ATR spectra recorded with a KRS-5 reflection element, at 45° angle of incidence. Times refer to chromic acid etch duration. (Ref. 6.)

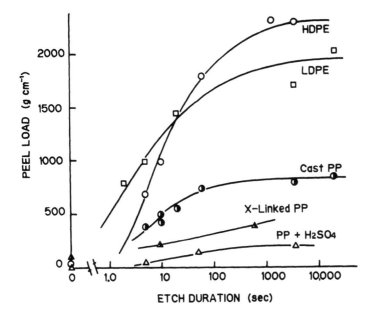

Figure 11.2   Peel loads for etched polyolefin film surfaces as a function of etch time.   (Ref. 6.)

through the identification of elements, and the combined state of an element can be determined from the chemical shift of the spectrum.

Figure 11.3 shows the results of polyethylene treated with Ar plasma for various periods of time [7].   It is clear that there is a spectrum due to an oxygen atom even for an untreated sample having zero time.   If the treating time is prolonged, the combined oxygen atoms increase and the shape of the carbon spectrum changes.   Nitrogen atoms do not appear before and after treatment. It is possible to know how carbon is combined with oxygen from the C_{1S} spectrum in this figure, especially from the skirts on the high-binding-energy side.

Figure 11.4 presents the results of theoretical calculations of 1S-electron binding energy changes when carbon, oxygen, or nitrogen combine with a variety of atoms [8].   Table 11.2 shows projections regarding carbon and oxygen bonds formed when high-density polyethylene (HDPE), polystyrene, polyethylene terephthalate (PET), polycarbonate, and so on, are treated with oxygen and hydrogen plasma [8].   This has revealed that a variety of functional groups are formed on the surface by plasma treatment.

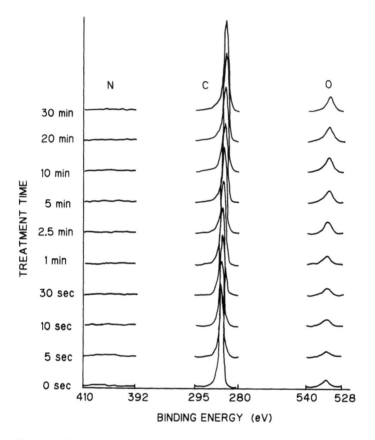

Figure 11.3   Relation between ESCA spectra and Ar$^+$ treatment time of PE.   (Ref. 7.)

However, ESCA measurement alone cannot achieve high relia-
bility in qualitative or quantitative analysis. Hence measures
capable of converting the functional groups into derivatives that
are capable of more accurate ESCA measurement have been tried.
Some examples are shown in Figures 11.5 [9] and 11.6 [10]. The
reason atoms such as F, Br, and Ag are introduced into the new
functional groups is that none of those atoms are contained in the
original materials near the surface, and hence are detectable by
ESCA with sufficient accuracy.

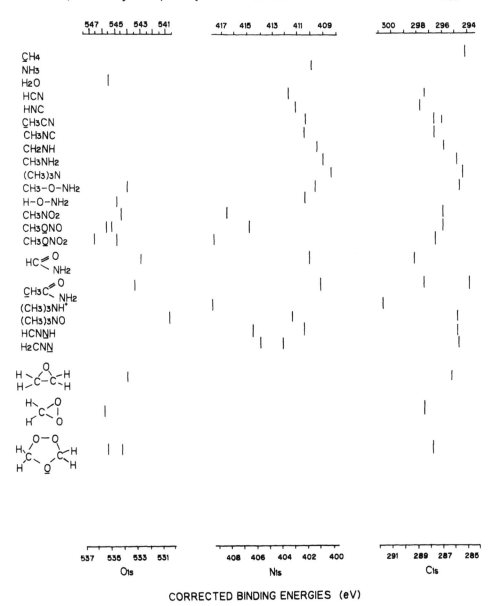

Figure 11.4 Theoretically computed (ΔSCF) calculation of C-1s, N-1s, and O-1s core-level binding energies. (Ref. 8.)

Table 11.2 Functional Groups of Various Polymer Surface by O_2 or H_2 Plasma Treatment (with ESCA)

Polymer	Plasma treatment	θ (deg)	C1S Total	$\underline{C}$—H	$\underline{C}$—O	$\underline{C}$=O	$O=\underline{C}$—O	$O-\underline{C}(=O)-C$	$\pi \to \pi^*$	C1S/O1S	C_{1S}^{30}/C_{1S}^{70}
HDPE	untreated	30	100	95.2	4.8	—	—	—	—	90.9	1.8
		70	100	93.9	5.6	0.5	—	—	—	90.9	
	O_2	30	100	84.4	10.1	4.4	1.7	0.8	—	5.4	1.6
		70	100	80.0	11.2	3.0	3.2	1.6	—	3.7	
	H_2	30	100	95.2	4.8	—	—	—	—	62.5	2.0
		70	100	92.5	6.4	1.2	—	—	—	62.5	
	H_2/O_2	30	100	80.0	9.7	4.8	4.1	1.4	—	3.8	1.8
		70	100	74.6	12.7	5.2	5.2	2.2	—	2.8	
	O_2/H_2	30	100	82.0	9.8	3.3	2.5	2.5	—	4.3	1.9
		70	100	78.1	12.5	4.7	3.9	0.8	—	3.4	
Polystyrene	untreated	30	100	90.1	4.5	—	—	—	5.4	52.6	1.9
		70	100	89.8	5.1	—	—	—	5.1	40.0	
	O_2	30	100	73.5	11.8	5.9	3.7	2.9	2.2	7.5	1.7
		70	100	61.3	15.3	11.0	7.4	4.3	0.6	1.5	
	H_2	30	100	84.8	12.1	1.5	—	—	1.5	11.2	2.0
		70	100	90.1	8.1	0.9	—	—	0.9	7.4	
	H_2/O_2	30	100	80.0	8.8	4.0	2.4	3.2	1.6	2.7	1.9

Material	Treatment										
PET	untreated	70	100	68.5	15.1	6.9	4.8	4.1	0.7	2.0	
	untreated	30	100	59.0	21.2	0.3	17.1	0.6	1.8	1.9	2.0
	O_2	70	100	61.3	17.2	4.9	14.1	0.6	1.8	2.2	
	O_2	30	100	52.4	25.7	—	19.9	1.6	0.5	1.5	2.7
	H_2	70	100	53.2	16.0	11.2	17.0	2.7	—	1.5	
	H_2	30	100	69.0	16.6	—	13.1	—	1.4	2.9	1.9
	H_2/O_2	70	100	78.2	14.1	—	7.7	—	—	3.8	
	H_2/O_2	30	100	53.3	22.6	0.7	19.0	2.9	1.5	1.6	2.3
	O_2/H_2	70	100	49.8	20.9	6.0	19.9	2.5	1.0	1.7	
	O_2/H_2	30	100	55.4	22.3	—	19.6	0.9	1.8	1.8	2.0
		70	100	54.3	24.5	—	19.6	1.1	0.5	1.8	
Polycarbonate	untreated	30	100	78.1	13.3	—	—	6.3	2.3	4.3	1.7
	untreated	70	100	78.4	13.5	—	—	6.1	2.0	4.1	
	O_2	30	100	66.7	16.7	5.3	4.0	5.3	2.0	2.6	1.4
	O_2	70	100	64.0	19.2	5.9	5.0	5.0	0.8	2.2	
	H_2	30	100	82.5	12.4	—	—	4.1	1.0	5.5	2.3
	H_2	70	100	86.1	10.8	—	—	2.3	0.7	5.2	
	H_2/O_2	30	100	69.4	17.6	3.5	2.4	5.9	1.2	2.6	1.9
	H_2/O_2	70	100	67.1	18.1	4.7	4.0	4.7	1.3	2.2	
	O_2/H_2	30	100	69.9	18.2	2.8	2.1	4.9	2.1	2.6	1.7
	O_2/H_2	70	100	69.8	17.4	4.7	3.5	4.7	—	2.4	

Source: Ref. 7,8.

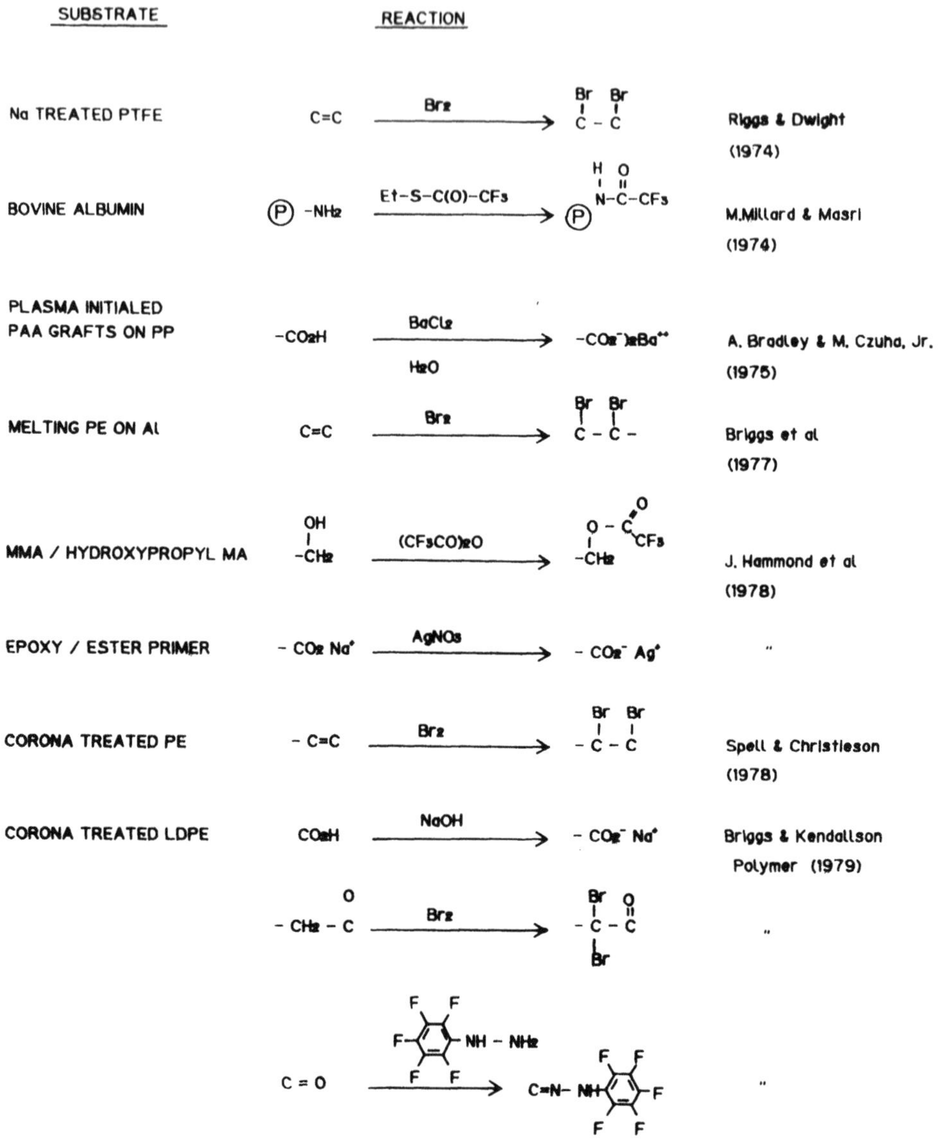

SUBSTRATE   REACTION

Na TREATED PTFE   C=C   —Br₂→   Br Br
                                           | | Riggs & Dwight
                                           C - C   (1974)

BOVINE ALBUMIN   Ⓟ -NH₂ —Et-S-C(O)-CF₃→ Ⓟ H O   M.Millard & Masri
                                               N-C-CF₃   (1974)

PLASMA INITIALED
PAA GRAFTS ON PP   -CO₂H —BaCl₂ / H₂O→ -CO₂⁻)₂Ba⁺⁺   A. Bradley & M. Czuha, Jr.
                                                          (1975)

MELTING PE ON Al   C=C   —Br₂→   Br Br
                                     | | Briggs et al
                                     C - C -   (1977)

MMA / HYDROXYPROPYL MA   OH
                         |
                        -CH₂ —(CF₃CO)₂O→   O - C⟨O / CF₃   J. Hammond et al
                                           -CH₂   (1978)

EPOXY / ESTER PRIMER   - CO₂ Na⁺ —AgNO₃→   - CO₂⁻ Ag⁺   "

CORONA TREATED PE   - C=C —Br₂→   Br Br
                                      | | Spell & Christieson
                                      - C - C   (1978)

CORONA TREATED LDPE   CO₂H —NaOH→   - CO₂⁻ Na⁺   Briggs & Kendallson
                                                          Polymer (1979)

                       O   Br O
                       || | ||
                    - CH₂ - C —Br₂→   - C - C   "
                                           |
                                           Br

                       F F
                    F⟨benzene⟩NH - NH₂   F F
          C = O ——————→   C=N- NH⟨benzene⟩F   "
                       F F   F F

Figure 11.5   XPS studies using polymer surface derivatization.
(Ref. 9.)

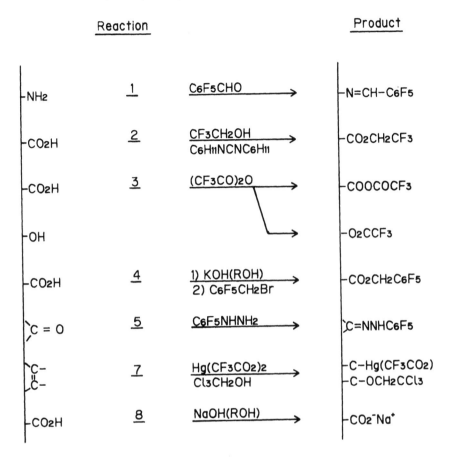

Figure 11.6   Relations of PE-Ar with various tagged reagents.
(Ref. 10.)

The surface of Teflon film (FEP) is chemically treated with, for example, an ammonium solution of sodium, to provide adhesive properties. Figure 11.7 presents structural changes in the film surface, treated as described above, examined by two methods, ESCA and the contact angle method [11].

The spectra of F_{1S} and C_{1S} ($-CF_2^*-$) are clearly shown in blank FEP (1), but when treated with Na/NH$_3$ (2) they disappear

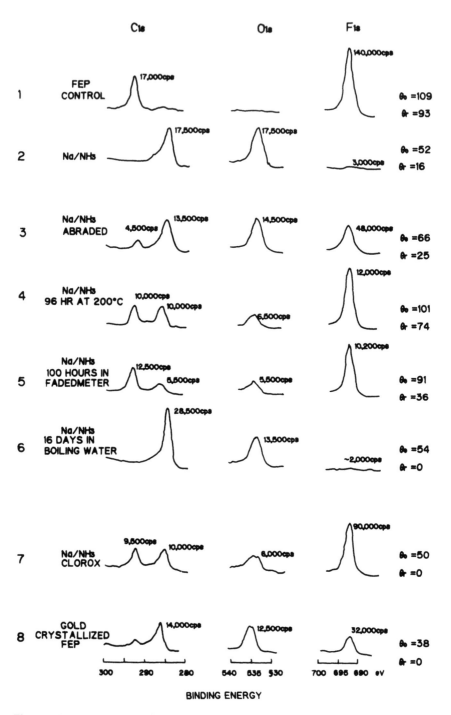

Figure 11.7   ESCA spectra and water contact angles from Teflon FEP before and after various surface treatments.   (Ref. 11.)

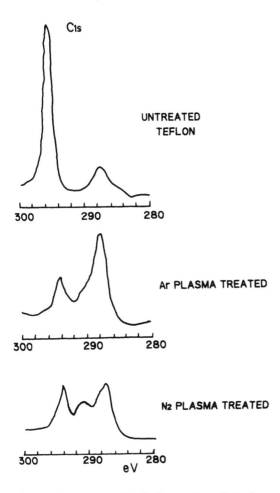

Figure 11.8   ESCA C-1s peaks of Teflon samples treated with Ar and N_2 plasma.   (Ref. 12.)

almost completely and then appear in the spectrum of a hydrocarbon having carbonyl and carboxyl groups. Because this change does not proceed at depths above 100 to 200 Å, its high sensitivity was made clear by ESCA and the reason for its good adhesiveness due to the Na/NH_3 treatment was also made clear.

The treatment of Teflon with Ar or N_2 plasma, as shown in Figure 11.8, also causes a decrease in the peak of C_{1S} ($-_*CF_2^*-$), accompanied by the appearance of new components due to $\overset{*}{C}-N$, $\overset{*}{C}-O$, and so on [12]. This suggests that in Ar plasma treatment, radicals are formed first and then oxygen atoms are captured by the etching, and that in N_2 plasma treatment, $C-N$ bonds are formed by direct reaction.

11.3 ANALYSES OF THE SURFACE AND THE INTERFACE OF FILLER

There are some hydroxyl groups on the surface of the filler of the glass fiber or oxide group [13]. In the case of carbon fiber, various functional groups are formed by various oxidation treatments [14]. It is well known that these functional groups are very useful in interactions with surface-treating agents such as silane coupling agents or with matrix resins.

The functional groups formed by oxidation treatment of carbon fiber, if detected by ESCA as shown in Figure 11.9 [15], are known to be mainly hydroxyl in the case of weak oxidation, and carboxyl in the case of strong oxidation: The results coincide with our results with ordinary chemical analysis [14].

To analyze these surface functional groups by separating with higher sensitivity, a method using the chemical modification reaction, which includes F in the reactant, is designed as shown in Figure 11.10 [16]. Two or three other examples [17] were examined in the same way by ESCA (XPS); the relation between the O/C ratio of the surface functional groups for various oxidation treatments and the shear strength of the composite material when manufactured are reported in Table 11.3 [1]. In another example the functional groups on the carbon black surface were measured by FT-IR and the carboxyl ($-COOH$) groups were quantitatively analyzed [18].

As described previously, there are silanol groups (Si$-$OH) on the surface of silica (SiO_2) and glass fiber. In one study these groups were analyzed quantitatively using ESCA by separating the peak of the O_{1S} spectrum. As shown in Figure 11.11 curve fitting resulting in a value of 532.4 eV for the chemical shift of O_{1S} and 533.3 eV for Si$-$OH makes it clear that the Si$-$OH groups decrease to about half at 500°C by heat treatment [19].

Because composite materials use these fillers after treatment of silane coupling agents, it is important to study the interaction

C1s difference spectra

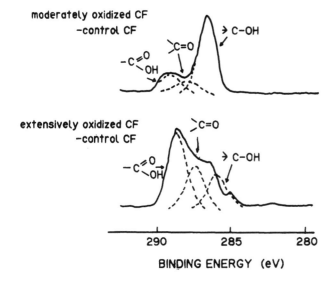

moderately oxidized CF
  −control CF

−C≮OH)C=O → C−OH

extensively oxidized CF
  −control CF

>C=O
−C≮OH → → C−OH

290   285   280

BINDING ENERGY (eV)

Figure 11.9   C-1s digital difference spectra for CF surface oxidation.   (Ref. 15.)

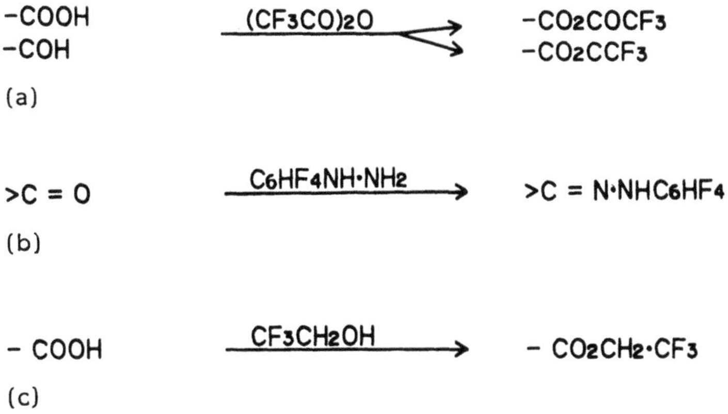

−COOH   (CF3CO)2O   −CO2COCF3
−COH   −CO2CCF3

(a)

>C = O   C6HF4NH·NH2   >C = N·NHC6HF4

(b)

− COOH   CF3CH2OH   − CO2CH2·CF3

(c)

Figure 11.10   Relations of carbon fiber surface with various
tagged reagents.   (Ref. 16.)

Table 11.3   Relation Between Functional Groups and Various
Surface Treatments of Carbon Fiber by XPS

| Surface treatment | O/C (%) | O_{1S} | | Shear strength (MPa) of epoxy composite |
		$\begin{matrix} -C-O^*H \\ \parallel \\ O \end{matrix}$	$\begin{matrix} -C-OH \\ \parallel \\ O^* \end{matrix}$	
None	4.3	—	—	34.6
Air oxidation (400°C, 1 h)	19.1	0.42	0.58	41.5
Chromic acid (15 min)	21.6	0.40	0.60	38.7
HNO_3 (3 h)	27.8	0.46	0.54	37.4
Hypochlorous acid (25 h)	22.3	0.37	0.63	46.2

Source: Ref. 1.

between the filler and the coupling agent.   There are many examples
of analyses that use FT-IR.   Figure 11.12 shows an analysis of
silica treated with a diaminosilane coupling agent [20].   In this
case *N*-2-aminoethyl-3-aminopropyltrimethoxysilane (AAPS) is adsorbed by interaction similar to a covalent bond.   This is caused
by the strong hydrogen bond between a silanol group and an amino
group on the surface of silica; the adsorption model is shown in
Figure 11.13 [20].

Glass fiber treated with silane coupling agents was analyzed
using ESCA; the structure around Si was analyzed using the Augur
parameter obtained from the peaks of Si_{1S} (E_B 1845 eV) and Si(KLL)
Augur (E_K 1605 eV), and from the energy difference between the
two.   These analyses have revealed (Figure 11.14) that silane molecules combine through treatment with silane-coupling agents, and
further, they react with the adjacent silanol group (Si—OH) to
form strong bonds [21].   Similar results obtained by analysis using
the Raman spectrum suggest that the action of the silane-coupling
agents is as follows [22]:

$$CH_2 {=} CH \; Si \; (O \cdot C_2H_5)_3 + 3H_2O \longrightarrow CH_2 = CH \; Si(OH)_3$$

$$+ 3 \; C_2H_5OH \qquad\qquad\qquad\qquad (1)$$

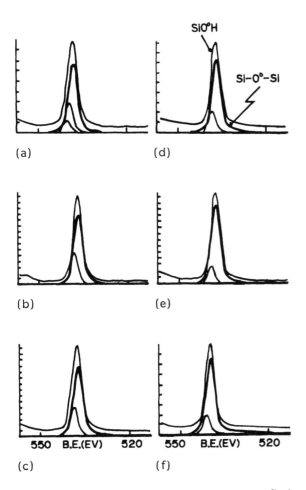

Figure 11.11   The O-1s spectral curve-fitting results for Cab-O-Sil silica at (a) −120°C, (b) 30°C, (c) 100°C, (d) 300°C, (e) 500°C, and (f) 700°C.   (Ref. 19.)

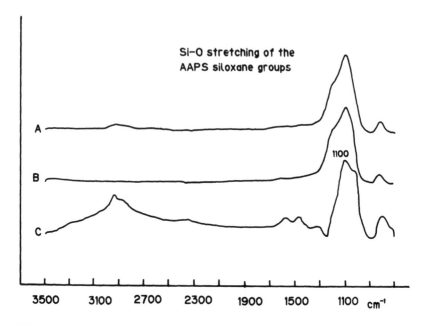

Figure 11.12   FT-IR absorbance spectra of diaminosilane coupling agent and Cab-O-Sil system.   A, Cab-O-Sil-treated 1% AAPS in toluene solution at 130°C; B, heat-cleaned Cab-O-Sil; C, difference spectrum of A − B.   (Ref. 20.)

A. DIAMINOSILANE
ON SILICA SURFACE

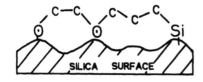

B. TRIAMINOSILANE
ON SILICA SURFACE

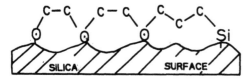

C: CH₂   ☉: NH or NH₂

Figure 11.13   Proposed structure of diaminosilane (AAPS) and tri-aminosilane adsorbed on silica surfaces.   (Ref. 20.)

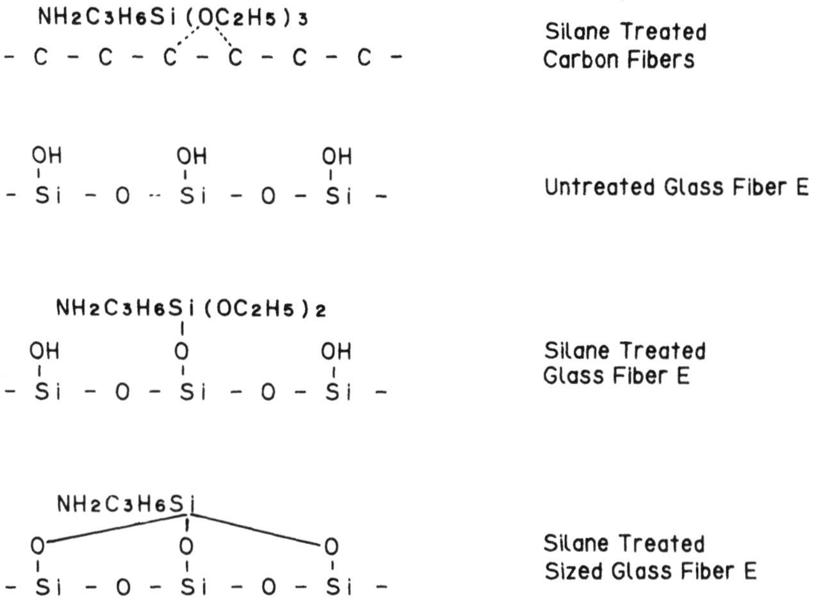

$NH_2C_3H_6Si(OC_2H_5)_3$

$- C - C - C - C - C - C -$ Silane Treated
Carbon Fibers

$$\begin{array}{cccc} OH & OH & OH \\ | & | & | \\ - Si - O - Si - O - Si - \end{array}$$ Untreated Glass Fiber E

$NH_2C_3H_6Si(OC_2H_5)_2$
$$\begin{array}{ccc} & | & \\ OH & O & OH \\ | & | & | \\ - Si - O - Si - O - Si - \end{array}$$ Silane Treated
Glass Fiber E

$NH_2C_3H_6Si$
$$\begin{array}{ccc} O & O & O \\ | & | & | \\ - Si - O - Si - O - Si - \end{array}$$ Silane Treated
Sized Glass Fiber E

Figure 11.14   Possible coupling of silane to carbon and glass fibers.   (Ref. 21.)

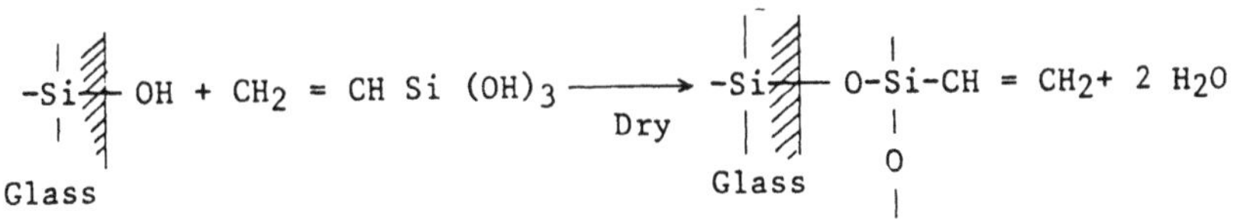

$$-Si\!\!-\!\!OH + CH_2 = CH\,Si\,(OH)_3 \xrightarrow{\text{Dry}} -Si\!\!-\!\!O-Si-CH = CH_2 + 2\,H_2O$$

Glass   Glass

(2)

11.4   ANALYSES OF THE INTERFACE BETWEEN FILLER AND MATRIX

It is possible to analyze and examine spectroscopically a broken-out section of composite material or a peeled adhesive surface.

11.4.1   Analysis Examples of Adhesion Interface

Figure 11.15 shows an ESCA projection of a Pd-PET adhesion interface of electrostatic printing material, made by spattering Pd thin film ($<$30 Å) and vapor deposition of selenium (Se) on the surface of polyethylene terephthalate (PET) [23].   Early in the spattering

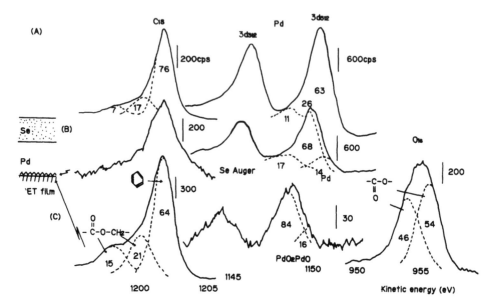

Figure 11.15   Adhesive interface between Pd and PET film.
(Ref. 23.)

period, Pd particles kick out oxygen from the PET surface and the
oxygen oxidizes Pd to form a layer of PdO or PdO_2. The metallic
Pd will adhere to the layer, and the layer gives the necessary
electroconductivity for electrostatic printing and good adhesion with
selenium metal. On the other hand, Pd oxide film will play the
role of intermediate between Pd metal and organic polymer and improve the adhesion.

11.4.2   Examples of Interface Between Silane-Treated Glass and Matrix Resin

Figure 11.16 shows the interface interaction between silane-treated
E glass and polystyrene examined by a Raman spectrum [24]. Comparison of B in the figure with C indicates that the polymerization
of styrene is proceeding on the silane-treated glass, and a comparison of C with D indicates that the interaction between the
silane-coupling agent and styrene and homopolymerization of the
styrene are taking place following the shift of absorption from
1718 cm^{-1} to 1702 cm^{-1}, as carbonyl stretching vibration of the
silane-coupling agent has revealed. Measurement by FT-IR [25] as
shown in Figure 11.17, and the observation of disappearing vinyl

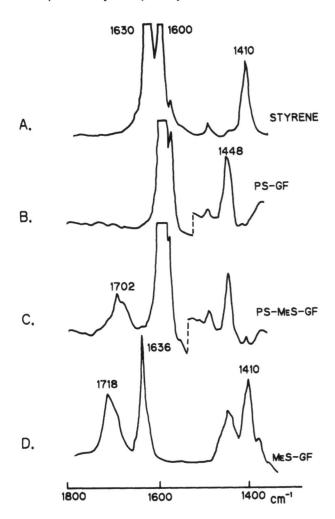

Figure 11.16   Raman spectra of a glass fiber/matrix interface. A, styrene monomer; B, untreated E-glass fiber coated with polystyrene; C, E-glass fiber treated with γ-methacryloxypropyltrimethoxysilane. Styrene monomer was polymerized on the glass surface; D, E-glass fiber treated with γ-methacryloxypropyltrimethoxysilane. (Ref. 24.)

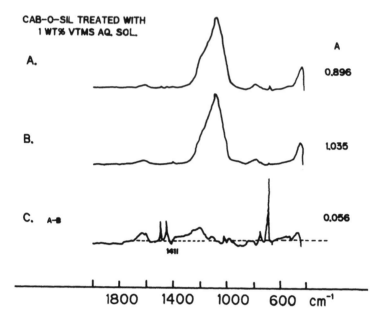

CAB–O–SIL TREATED WITH
1 WT% VTMS AQ. SOL.

A.

A
0.896

B.

1.035

C. A–B

0.056

1411

1800   1400   1000   600   cm⁻¹

Figure 11.17   High-surface-area silica treated with aqueous solution of 1 wt % vinyltrimethoxysilane (VTMS). A, silica was polymerized with styrene and washed with CS_2 three times; B, polystyrene produced in experiment A was deposited on the silica and then
the silica was washed with CS_2 three times; C, difference spectrum
of A − B. (Ref. 25.)

groups in the silane-coupling agent and of the formation of polystyrene on the silica, have confirmed the occurrence of a reaction
between the polymer and the silane-coupling agent.

Furthermore, observation of the spectrum for styrene polymerized
on the surface of silane-treated silica and of the difference spectrum
of polystyrene adsorbed on the surface of silica have revealed that
there are absorption bands of attactic polystyrene at 1602, 1493,
1453, 756, and 698 cm⁻¹: 1411 and 1010 cm⁻¹ are absorption bands
related to vinyltrimethoxysilane, and C of the difference spectrum
is below the baseline; this indicates that the vinyl groups of silane
react with styrene to form a copolymer.

11.4.3   Analysis of Interface Between Fiber and Metallic Matrix

Analysis by SIMS (secondary ion mass spectrometry) of the interface in the carbon fiber/Al composite material is discussed next.

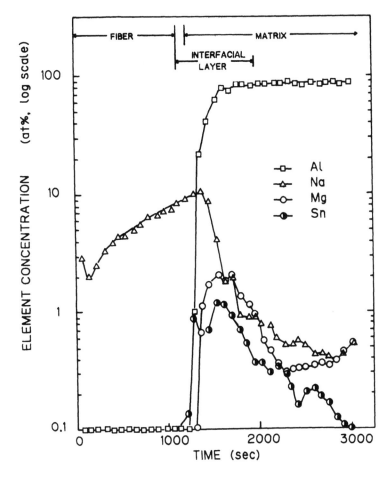

Figure 11.18   IMMA plot of sodium, tin, magnesium, and aluminum concentrations as a function of time (depth of sputtered hole) in graphite-aluminum composite prepared by the sodium process. (Analysis starts in graphite fiber and proceeds into the aluminum matrix.)   (Ref. 26.)

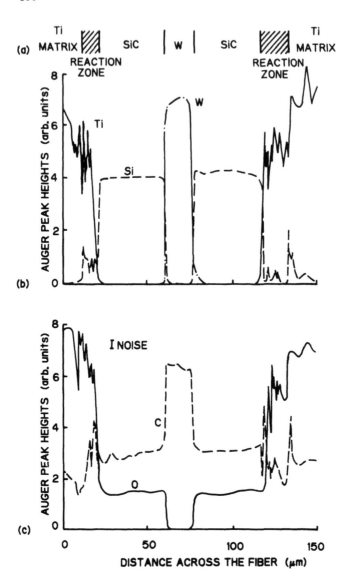

Figure 11.19 Overview line scans of the composite sample: (a) schematic diagram of the sample (the shaded regions represent the reaction zone); (b) Ti, Si, and W line-scan profiles; (c) C and O line-scan profiles. The maximum FE noise is indicated by an error bar. (Ref. 28.)

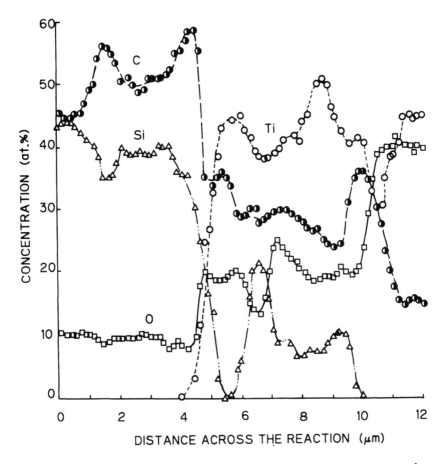

Figure 11.20 Detailed analysis of the reaction zone: ◑, carbon; △, silicon; ○, titanium; □, oxygen. (Ref. 28.)

In this system, a sharp decrease in strength arises from the formation of Al_4C_3 on the interface initiated at about 773 K (500°C). To
suppress such a reaction and to improve the wettability of Al, pretreatment of the carbon fiber is necessary; in this case the fiber
is treated in a fused metallic bath of Na and Sn (a small amount of
Mg is added).

Figure 11.18 shows measurements by SIMS which indicate an
interface reaction phase consisting of Al, Na, Mg, and Sn [26].
The wettability of carbon fiber to Na is good, resulting in good
penetration of Na into the carbon fiber; this penetrating phenomenon
is believed to arise from intergranular diffusion or the formation of
an intercalation compound such as $C_{64}Na$. Carbon fiber is first
covered by Na-Sn intermetallic compound, and then by Mg-Sn intermetallic compound: This phenomenon is supposed to be due to the
geometric effect of the irregular surface of carbon fiber or to a
decrease of concentration on the interface. The thickness of the
interface reaction phase is 0.45 to 0.55 μm.

There are other examples of analyses by HEED (high-energy
electron diffraction) and SEM (scanning electron microscopy) [27].
Figure 11.19 shows an analysis by AES (Auger electron spectroscopy) of the interface of a titanium composite material reinforced
with W-SiC fiber [28]. The cross section with a fiber in its center
is line-analyzed with a beam diameter of less than 50 nm. The
results have revealed that a reaction phase of 7 to 12 μm width is
formed on the interface between the Sic fiber and the Ti matrix.

Figure 11.20 presents a detailed analysis of the reaction phase
[28]. These results indicate that an Si element exists on the outside of the reaction phase and that W is transformed into WC and
Ti matrix into its oxide. Because there is TiC in the reaction phase
and C in the Ti matrix, C of SiC is supposed to have moved to the
outside. Similar studies have been reported [29,30].

REFERENCES

1.   K. Kubo, M. Koishi, and T. Tsunoda, Eds., *Composite Materials and Interface*, Sōgō Gijutsu Shuppan, Tokyo, 1986, p. 13.

2a. K. Noma and R. Yosomiya, A Discussion on the Surface of Resin by IR Analysis, *Resin Finish. Jpn.*, *12*, 35 (1963).

2b. J. Shimada and M. Hishino, Surface Fluorination of Transparent Polymer Film, *J. Appl. Polym. Sci.*, *19*, 1439 (1975).

3.   P. Blais, M. Day, and D. M. Wiles, Photochemical Degradation of Polyethylene Terephthalate, *J. Appl. Polym. Sci.*, *17*, 1895 (1973).

4.   J. P. Luongo and H. Schonhorn, Infrared Study of Substrate Effects in the Surface Region of Polyethylene, *J. Polym. Sci. A-2*, *6*, 1649 (1968).

5a. W. T. M. Johnson, The Chemical Nature of Paint Film Surface, *Off. Dig. Oil Colour Chem. Assoc.*, *32*, 1067 (1960).

5b. W. T. M. Johnson, Surface Analysis and Adhesion, *Off. Dig. Oil Colour Chem. Assoc.*, 33, 1489 (1961).

6. P. Blais, D. J. Carlsson, G. W. Csullog, and D. M. Wiles, The Chromic Acid Ething of Polyolefin Surfaces and Adhesive Bonding, *J. Colloid Interface Sci.*, *47*, 636 (1974).

7. Y. Ikada, Three Topics on Polymer Surfaces, *Surf. Jpn.*, *22*, 119 (1984).

8. D. T. Clark, The Modification, Degradation and Synthesis of Polymer Surfaces Studied by ESCA, *ACS Symp. Ser.*, *162*, 247 (1981).

9. C. D. Batich and R. C. Wendt, Chemical Labels to Distinguish Surface Functional Groups Using X-Ray Photoelectron Spectroscopy (ESCA), *ACS Symp. Ser.*, *162*, 221 (1981).

10. D. S. Everhart and C. N. Reilly, Chemical Derivatization in Electron Spectroscopy for Chemical Analysis of Surface Functional Groups Introduced on Low-Density Polyethylene Film, *Anal. Chem.*, *53*, 665 (1981).

11. W. M. Riggs and D. W. Dwight, Characterization of Fluoropolymer Surfaces, *J. Electron Spectrosc. Relat. Phenom.*, *5*, 447 (1974).

12. H. Yasuda, Plasma for Modification of Polymer, *J. Macromol. Soc. Chem. Ed.*, *A-10*, 383 (1976).

13. K. Hashimoto, T. Fujisawa, M. Kobayashi, and R. Yosomiya, Graft Copolymerization of Glass Fibers and Its Application, *J. Appl. Polym. Sci.*, *27*, 4529 (1982).

14. A. Tanaka, T. Fujisawa, and R. Yosomiya, Graft Polymerization of Carbon Fiber, *J. Polym. Sci. Polym. Chem. Ed.*, *18*, 2267 (1980).

15. T. Takahagi and A. Ishitani, XPS Studies by Use of the Digital Difference Spectrum Technique of Functional Groups on the Surface of Carbon Fiber, *Carbon*, *22*, 43 (1984).

16. K. Morita, Y. Murata, A. Ishitani, K. Murayama, T. Ono, and A. Nakajima, Characterization of Commercially Available PAN Based Carbon Fibers, *Pure Appl. Chem.*, *58*, 455 (1986).

17. A. Proctor and P. M. A. Sherwood, X-Ray Photoelectron Spectroscopic Studies of Carbon Fiber Surfaces II, *Carbon*, *21*, 53 (1983).

18. J. M. O'Reilly and R. A. Mosher, Functional Groups in Carbon Black by FTIR Spectroscopy, *Carbon*, *21*, 47 (1983).

19. M. L. Miller and R. W. Linton, X-Ray Photoelectron Spectro-
     scopy of Thermally Treated SiO_2 Surfaces, *Anal. Chem.*, *57*,
     2314 (1985).

20. C. H. Chiang and J. L. Koenig, Fourier Transform Infrared
     Spectroscopic Study of the Adsorption of Multiple Amino
     Silane Coupling Agents on Glass Surfaces, *J. Colloid Inter-
     face Sci.*, *83*, 361 (1981).

21. K. Yates and R. H. West, Monochromatized Ag Lα X-Rays as
     a Source for Higher Energy XPS, *Surf. Interface Anal.*, *5*,
     133 (1983).

22. J. L. Koenig and P. T. K. Shih, Raman Studies of the Glass
     Fiber-Silane-Resin Interface, *J. Colloid. Interface Sci.*, *36*,
     247 (1971).

23. A. Ishitani, Surface Analysis of Polymers by ESCA, *Surf.
     Jpn.*, *17*, 26 (1979).

24. H. Ishida and J. L. Koenig, The Reinforcement Mechanism of
     Fiber Glass Reinforced Plastics Under Wet Conditions, *Polym.
     Eng. Sci.*, *18*, 128 (1978).

25. H. Ishida and J. L. Koenig, An Investigation of the Coupling
     Agent/Matrix Interface of Fiber Glass Reinforced Plastics by
     Fourier Transform Infrared Spectroscopy, *J. Polym. Sci.
     Polym. Phys. Ed.*, *17*, 615 (1979).

26. D. M. Goddard, Interface Reactions During Preparation of
     Aluminum-Matrix Composites by the Sodium Process, *J.
     Mater. Sci.*, *13*, 1841 (1978).

27. I. H. Khan, The Effect of Thermal Exposure on the Mechan-
     ical Properties of Aluminum-Graphite Composites, *Met. Trans.*,
     7A, 1281 (1976).

28. E. P. Zironi and H. Poppa, Micro-area Auger Analysis of a
     SiC/Ti Fiber Composite, *J. Mater. Sci.*, *16*, 3115 (1981).

29. R. Browning, Ratioed Scatter Diagram: An Erotetic Method for
     Phase Identification on Complex Surface Using Scanning Auger
     Microscopy, *J. Vac. Sci. Technol.*, *2A*, 1453 (19??).

30. V. W. Bermudez, Auger and Electron Energy-Loss Study of
     the Pd/SiC Interface and Its Dependence on Oxidation, *Appl.
     Surf. Sci.*, *17*, 12 (1983).

12

Interfacial Strength of Composite Materials

12.1   INTRODUCTION

Research on the strength of fiber-reinforced plastics (FRP) is
roughly classified into two approaches: macroscopic and microscopic.
The former, called the "parallel spring model," has long been used.
In this method, a simple model composed of a fiber and a matrix is
substituted for a FRP structure, and such factors as elastic coefficient, strength, and so on, are calculated out on the model under
the presumption that the longitudinal strains of both structures are
the same.

On the other hand, in the microscopic approach the intent is to
observe the structural behavior from the distribution of inner stress
and strain of the elements of FRP, and accordingly, to seek out
"knee phenomena" appearing in a stress-strain curve, the formation
of cracks, crack propagation, and so on. Methods for such research
analysis are studied from various angles, such as simple analysis
employing a complex principle, analysis by a theory of orthogonal
anisotropy, analysis by an energy method, and elastoplastic analysis
in the case of a discontinuous fiber-reinforcing material.

FRP is a composite material made by combining materials of differing composition. It is therefore natural that the nature of the
interface between the materials has so great an effect on the properties of FRP. The interface between the materials is completely
different from the nature of the materials, and it is extremely difficult at present to make a quantitative determination of the mechanical properties of this interfacial portion.

In this chapter we deal with the results of evaluation of stress
that occurs at the interfacial portion between the fiber and the
matrix in a stressed FRP and the interfacial strength of that portion.

12.2   BASIC STRENGTH OF FIBER-REINFORCED COMPOSITES

12.2.1   Strength of Continuous Fiber-Reinforced Composite Material

If the direction of tensile load is parallel to that of the fiber axis and the homogeneously strained condition is maintained until the moment of fracture of the composite material, the breaking stress of fiber (ε_{fu}), the breaking stress of matrix (ε_{mu}), and the mean strain of the composite material (ε_c) satisfy the following equation [1]:

$$\varepsilon_c = \varepsilon_{fu} = \varepsilon_{mu} \qquad \varepsilon_{fu} \leqslant \varepsilon_{mu} \tag{1}$$

thus

$$\frac{\sigma_f}{E_f} = \frac{\sigma_m}{E_m} \qquad \sigma_{fu} \leqslant \frac{E_f}{E_m} \sigma_{mu} \tag{2}$$

where σ_f, E_f, σ_m, and E_m are, respectively, the strength of the fiber and its elastic modulus, and the strength of the matrix and its elastic modulus.   The stress applied to the composite material is

$$\sigma_c = V_f \sigma_f + (1 - V_f)\sigma_m \tag{3}$$

Therefore, the strength of the composite material (X) and the equivalent fiber stress ($\bar{\sigma}_{fu}$) are given by

$$\frac{X}{V_f} = \bar{\sigma}_{fu} = \sigma_{fu}\left(1 + \frac{1 - V_f}{V_f}\frac{E_m}{E_f}\right) \tag{4}$$

where V_f is the volume content of the fiber.
    The fiber content V_{fcr} at which X becomes equal to the strength of matrix is given as

$$V_{fcr} = \frac{\sigma_{mu} - (\sigma_m)\varepsilon_{fu}}{\sigma_{fu} - (\sigma_m)\varepsilon_{fu}} \tag{5}$$

The fiber content ($V_{f,min}$) at which the fracture of fiber and the fracture of matrix take place at the same time can be given as

$$V_{f,min} = \frac{\sigma_{mu} - (\sigma_m)\varepsilon_{fu}}{\sigma_{fu} - \sigma_{mu} - (\sigma_m)\varepsilon_{fu}} \tag{6}$$

Figure 12.1 shows a relationship between σ_c and V_f, and examples of V_{fcr} and $V_{f,min}$ [2]. For composite material reinforced by the filament winding method, the experimental values are in fair agreement with the calculated values. However, for composite material reinforced with glass fiber satin weave fabric, a considerable difference is observed between these values. Because the content of the fibers parallel to the direction to which the stress is applied is smaller than the apparent fiber content, the strength of the composite material is usually represented approximately by multiplying the apparent fiber content by the reducing coefficient β, as in the following equation:

$$\frac{X}{V_f} = \sigma_{fu}\left(\beta + \frac{1 - V_f}{V_f}\frac{E_m}{E_f}\right) \tag{7}$$

where the value of β is about 0.5 in case of a fabric and about 3/8 in case of a mat.

The theory described above is a so-called phenomenalistic one which explains the strength characteristics by the basic values based on macroscopical experimental facts in a particular direction when various loads are applied to the unidirectional reinforcing material. Although this theory is highly practical, the influence exerted on the macroscopic strength of the composite material by the interfacial strength, which is the accumulation of the microscopic characteristics, has not been sufficiently elucidated. That is, the problem is extremely difficult because both the geometric random tendency of the fibers dispersed in the matrix resin and the probability property included in the strength problem itself are complicated.

A few reports suggest the relation of the interfacial strength to the shear fracture of the actual composite material. As dynamic models, a hexagonal arrangement (a) and a square arrangement (b) in which the fibers are regularly placed in order are used, as shown in Figure 12.2 [3a]. In these cases, the fibers are considered to be infinitely long in the direction of the Z axis. The critical value of the macroscopic stress acting from any direction except in the direction of the Z axis, the so-called off-axis strength, is determined numerically by these models, as is the relationship between the off-axis strength and the interfacial strength.

That is, to calculate the response to a force that acts macroscopically from any arbitrary direction, the outer load is decomposed by the rotation of the axis of coordinates, as shown in Figure 12.3 [3b]. For example, when a uniaxial load $\bar{\sigma}$ in the direction of the Z'' axis acts, the following equations are given:

$$\bar{\sigma}_x = \ell_\theta^2 m_\phi^2 \bar{\sigma}$$

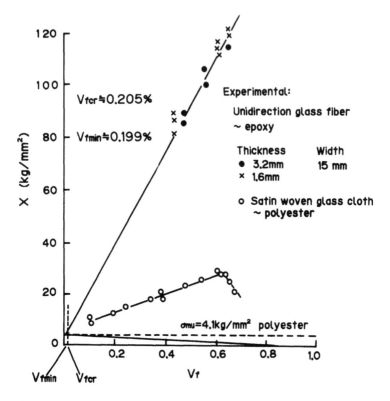

Figure 12.1   Fundamental strength (X) of continuous fiber-reinforced plastics.   (Ref. 2.)

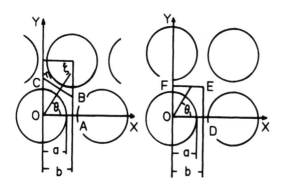

Hexagonal Array   Square Array

Figure 12.2   Patterns of fiber packing.   (Ref. 3a.)

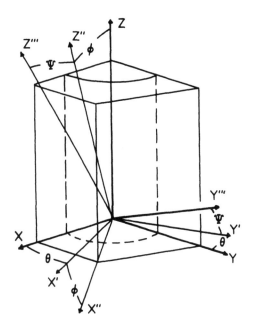

Figure 12.3   Alteration for the axis of coordinate.   (Ref. 3b.)

$$\bar{\sigma}_y = m_\theta^2 m_\phi^2 \bar{\sigma}$$

$$\bar{\sigma}_z = \ell_\phi^2 \bar{\sigma}$$

$$\bar{\tau}_{xz} = \ell_\theta \ell_\phi m_\phi \bar{\sigma} \qquad\qquad (8)$$

$$\bar{\tau}_{yz} = m_\theta \ell_\phi m_\phi \bar{\sigma}$$

$$\bar{\tau}_{xy} = \ell_\theta m_\theta m_\phi^2 \bar{\sigma}$$

where $\ell_\theta = \cos\theta$, $m_\theta = \sin\theta$, $\ell_\phi = \cos\phi$, and $m_\phi = \sin\phi$. For the six stress components thus determined, the respective breaking standards in the fiber and matrix are applied.

With respect to the standard in the fiber, Hoffman's condition shown by the following equation is applied:

$$C_1[(\sigma_y - \sigma_z)^2 + (\sigma_z - \sigma_x)^2] + C_3(\sigma_x - \sigma_y)^2 + C_6\sigma_z$$

$$+ C_7(\tau_{yz}^2 + \tau_{zx}^2) + 2(C_1 + 2C_3)\tau_{xy}^2 = 1 \qquad (9)$$

For the standard in the matrix resin, the maximum main stress condition is applied. In this case, the main stress σ_i of each component is determined by solving the following characteristic equation from the six stress components:

$$|\sigma_{ij} - \sigma\delta_{ij}| = 0 \qquad (10)$$

where δ_{ij} is Kronecker's delta. If the main stress σ_i is in the range from the compressive strength σ_{cc} to the tensile strength σ_{TC} of the resin, that is,

$$\sigma_{cc} \leq \sigma_i \leq \sigma_{TC} \qquad (11)$$

the fracture is regarded as not taking place.

Then the interfacial strength standard between the fiber and the matrix is determined. As shown schematically in Figure 12.4, the values considered are σ_{rc}^{IF}, which is the critical vertical stress at the interface; $\tau_{r\theta c}^{IF}$ and τ_{rzc}^{IF}, which are the critical shear stresses in the $r\theta$ and rz directions at the interface, respectively (shown by a square on the interface in the figure); and τ_{rc}^{IF}, which is obtained by the synthesis of these (shown by a dashed-line circle on the interface in the figure) [3b]. Of these, σ_{rc}^{IF} is regarded as the positive value, that is, the critical value of the tensile stress.

By using each basic value of the unidirectional carbon-fiber-reinforced epoxy resin, the relationship between the critical vertical tensile strength of the interface and the off-axis tensile strength of the unidirectional material was determined [4], with the results shown in Figure 12.5. The flat area in the figure is the region where the interfacial strength is not a determining factor, but where the maximum main stress in the resin becomes critical $(\sigma_{TC})_m$.

From this fact it can be seen that the influence of the vertical strength at the interface is minimal when the off-axis angle is small, and that the region where the strength is dominant broadens as the angle approaches 90°, that is, at a right angle to the fiber. When ϕ is 90°, the interfacial strength is a little lower than the determined maximum resin main stress of 10 kg/mm^2 and thus ceases to act at the determining factor. This difference is considered to result from the residual stress.

Figure 12.6 shows a relationship between the interfacial shear strength and the macroscopical shear strength, where the

Matrix

Fiber-Matrix Interface

Fiber

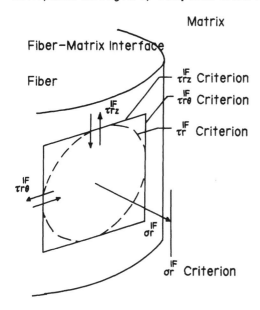

τ_{rz}^{IF} Criterion

$\tau_{r\theta}^{IF}$ Criterion

τ_{r}^{IF} Criterion

σ_{r}^{IF} Criterion

Figure 12.4 Schematic representation of the interfacial strength standard between fiber and matrix. (Ref. 3b.)

macroscopical shear strength is the reverse shear of the off-axis [4]. When ϕ is 0°, the genuine shear strength along the fiber of the unidirectional material is shown. Also, in this figure, the criticality transition takes place where the interfacial shear strength crosses the allowable main stress of the resin.

12.2.2 Strength of Discontinuous Fiber-Reinforced Composites

When the resin is reinforced with discontinuous fibers, the transmission of the load is carried out through the interface between the fibers and the matrix. Therefore, the interfacial characteristics between the dispersing phase (fibers) and the matrix, the ratio of the diameter to the length of the fiber, the ratio of the elastic modulus of the fiber to that of the matrix, and similar factors become important. The discontinuous fibers are actually irregularly dispersed in the matrix. Nevertheless, in most cases the analyses are carried out under the assumption that in general the fibers are unidirectionally oriented and approximately homogeneously dispersed.

For example, Figure 12.7 shows the Cox model [5]. Cox derived an equation based on the following assumptions: that an elongated fiber with a length of ℓ was buried in an elastic matrix,

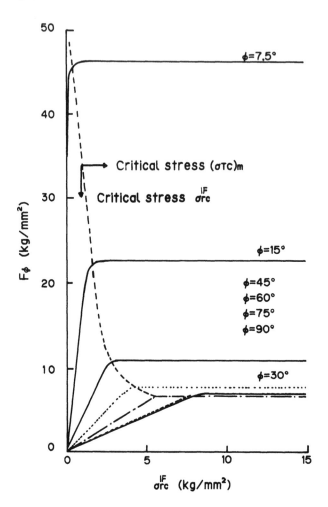

Figure 12.5   Relation between the critical vertical tensile strength
of the interface and the off-axis tensile strength of the unidirectional
material.   (Ref. 4.)

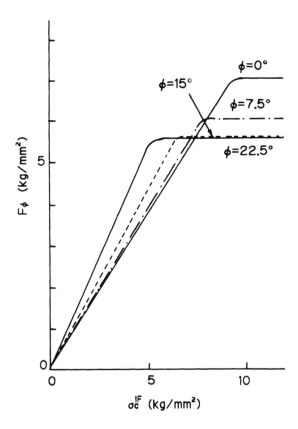

Figure 12.6   Relation between the interfacial shear strength and the shear strength of the unidirectional material.   (Ref. 4.)

that the fiber adhered completely to the matrix, that the strain of the fiber was equal to that of the interface of the matrix when stress was exerted in the direction of the fiber axis, and that there was no transmission of the stress from the cross section of the fiber:

$$\sigma_f = \frac{E_f - E_m}{E_m} \, \sigma_c \left[1 - \frac{\cosh \, \beta(\ell/2 - x)}{\cosh \, \beta\ell/2} \right] \tag{12}$$

Hence, with $x = \ell/2$, the maximum stress is given by

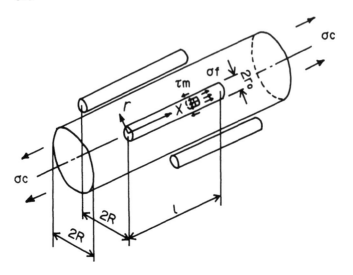

Figure 12.7   Model of Cox.   (Ref. 5a, p. 97)

$$\sigma_{fmax} = (E_f - E_m)\left(1 - \text{sech}\ \frac{\beta \ell}{2}\ \varepsilon\right)$$

$$\varepsilon = \frac{\sigma_c}{E_m} \tag{13}$$

where E_f and E_m are the elastic modulus of the fiber and that of
the matrix, respectively, and σ_c and σ_f are the stress of the matrix
and that of the fiber in the longitudinal direction, respectively.
   If $\ell = \ell_c$ and $\sigma_{fmax} = \varepsilon E_f$, ℓ_c is represented by

$$\ell_c = \frac{2}{\beta}\ \cosh^{-1}\left(\frac{E_f}{E_f - E_m}\right) \tag{14}$$

This equation indicates that when ℓ is larger than ℓ_c, the fiber
reaches its critical stress and a reinforcing effect can be exerted.
Because of this, ℓ_c is referred to as the critical fiber length.

$$\tau_m = \frac{(E_f - E_m)r_f\beta\ \sinh\ \beta(\ell/2 - x)}{2\ \cosh\ \beta(\ell/2)}$$

$$\beta = \sqrt{\frac{G_m}{E_r}\ \frac{2\pi}{A_f\ \ln(R/r_0)}} \tag{15}$$

where r_0 is the radius of the cross section of the fiber, 2R the distance between the centers of the cross sections of the fibers adjacent to each other, and A_f the cross-sectional area.

In addition, there are models by Outwater [6], Dow [7], Rosen [8], and others. With respect to interfacial shear stress, however, the experimental values determined by the photoelastic experiment are considerably different from the values obtained from these theoretical equations, as shown in Figure 12.8 [9]. This results from the formation of stress concentration at the ends of fiber, as apparent from the distribution of the main stress lines around the fiber shown in Figure 12.9 [9]. It is the main reason that the deformation of the matrix is restricted by fibers with high rigidity.

Next we describe the Kelly-Tyson equation [11], which is said to be simple in its expression and to represent the actual conditions reasonably well. If the tensile stress is given along the fiber axis, that is, in the x direction, shown in Figure 12.7 from the point of infinity, the shear stress acts on the side of the fiber. When the minute length of the fiber is given as dx, the variation in tensile load dP_f exerted on the fiber is equal to the shear force at the interface in the region of dx. Therefore, when the shear stress is given as τ_m, the following equation is obtained:

$$\frac{dP_f}{dx} = 2\pi r_f \tau_m \tag{16}$$

According to the Kelly-Tyson equation, the portions in the neighborhood of the fiber ends that have high interfacial shear stress are in the plastic region, and τ_m is assumed to reach the constant value. Hence, from equation (16), the following equation is derived:

$$P_f = 2\pi r_f \tau_m x \tag{17}$$

The stress that acts on the cross section of the fiber is

$$\sigma_f = \frac{2\tau_m}{r_f} x \tag{18}$$

σ_f increases linearly as the position on which it acts goes away from the fiber ends. When the strain ε_f of the fiber to this stress becomes equal to the strain ε_m of the matrix, the interfacial shear stress ceases to act and σ_f reaches a constant value. That is, the stress distribution shown in Figure 12.10 is obtained. When the maximum value σ_{fmax} of σ_f reaches the tensile strength σ_{fu} of the fiber, the fiber begins to break. If the length of the fiber at this time is given as ℓ_c, the critical fiber length, the following equation is obtained:

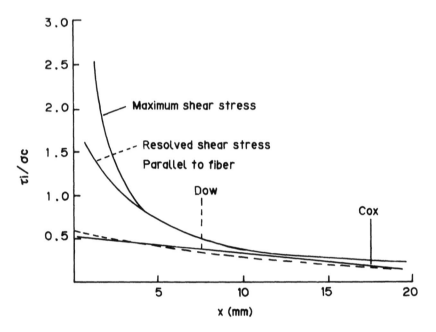

Figure 12.8   Comparison of experimental results with theoretical
curves for τ_i/σ_c at a distance x from the fiber end.   (Ref. 9.)

$$\frac{1}{2}\ell_c = \frac{r_f}{2}\frac{\sigma_{fu}}{\tau_m} \tag{19}$$

and the critical aspect ratio is given by

$$\frac{\ell_c}{d_f} = \frac{\sigma_{fu}}{2\tau_m} \tag{20}$$

If the aspect ratio ℓ/d_f of the fiber is higher than this value, the
fiber shows a reinforcing effect.

Published reports dealing with elastoplastic deformation caused
by interfacial stress between the fiber and the matrix are limited [12].
Next, a model is considered in which fiber of length ℓ and radius
r is burried in a cylindrical completely elastoplastic-body matrix of
radius R, as shown in Figure 12.11 [13].   The content by
volume of this fiber is represented by $(r/R)^2 = v$.   When the
average deformations of the fiber and the matrix are given as $\overline{U}_F$
and $\overline{U}_M$, respectively, and the interfacial deformations of the fiber

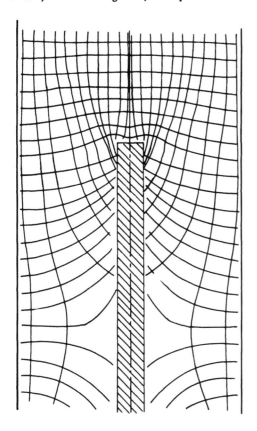

Figure 12.9   Tensile stress trajectories derived from the isoclinic fringe pattern.   (Ref. 9.)

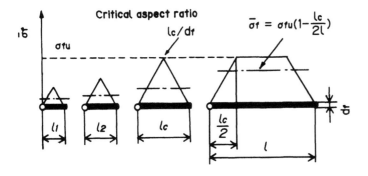

Figure 12.10   Average tensile stress for various fiber lengths. (Ref. 11.)

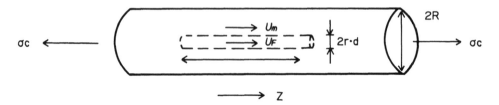

Figure 12.11   Characteristic volume element of composite.   (Ref. 13.)

and the matrix are given as U_F and U_M, respectively, the strains are given by

$$\bar{\varepsilon}_F = \frac{\partial \bar{U}_P}{\partial z} \qquad \bar{\varepsilon}_M = \frac{\partial \bar{U}_M}{\partial z} \qquad \gamma_F = \frac{U_P - \bar{U}_P}{h_F} \qquad \gamma_M = \frac{\bar{U}_M - U_M}{h_M}$$

$$(21)$$

where ε and γ are the vertical and shear strains, respectively, h_F and h_M are constants determined depending on r and R, and the equations corresponding to the various fiber arrangements are derived [1].

Considering the balance in the direction of the fiber axis, the following equation is provided:

$$\frac{\partial}{\partial z}\,\bar{\sigma}_F = -\frac{2}{r}\,\tau, \quad \frac{\partial}{\partial z}\,\bar{\sigma}_M = \frac{2r}{R^2 - r^2}\,\tau = \frac{2}{r}\frac{v}{1-v}\,r \qquad (22)$$

Further, considering the rule of mixture, the following equation is given:

$$\sigma_c = \bar{\sigma}(z) = v\bar{\sigma}_F(z) + (1 - v)\bar{\sigma}_M(z) \qquad (23)$$

If the stress does not act on the fiber ends, the following equation is obtained:

$$\bar{\sigma}_F\left(\pm\frac{\ell}{2}\right) = 0 \qquad (24)$$

When the fiber adheres completely to the matrix (i.e., $U_F = U_M$), the interfacial shear stress is given by

$$\tau = \frac{G}{h}\,(\bar{U}_m - \bar{U}_F) \qquad \frac{G}{h} = \frac{G_F G_M}{G_F h_M + G_M h_F} \qquad (25)$$

The elastic rule regarding the fiber and the matrix is as follows:

$$\bar{\sigma}_F = E_F \bar{\varepsilon}_F \qquad \tau_F = G_F \gamma_F \qquad \bar{\sigma}_M = E_M \bar{\varepsilon}_M \qquad \tau_M = G_M \gamma_M \qquad (26)$$

By solving the equation (22) under the boundary conditions of equations (23) and (24), the following equation is obtained:

$$\bar{\sigma}_F = \sigma_c \frac{E_F}{\bar{E}}\left(1 - \frac{\cosh Kz}{\cosh K\ell/2}\right)$$

$$\tau = \frac{K\gamma}{2}\,\sigma_c \frac{E_F}{\bar{E}}\,\frac{\cosh Kz}{\cosh K\ell/2} \qquad (27)$$

where

$$K = \frac{2}{r_h}\,\frac{G\bar{E}}{(1 - v)E_M E_F} \qquad \bar{E} \equiv VE_F - (1 - v)E_M \qquad (28)$$

From equation (27), in the case of a sufficiently long fiber, the tensile stress is $\sigma_c E_F/\bar{E}$ at a large portion of the fiber, and decreases to zero at a position $1/K$ away from the fiber ends. The interfacial shear stress is nearly zero beyond the range of a distance of $1/K$ from both ends of the fiber. Its value is maximal at a position $1/K$ away from both ends of the fiber and is given by

$$\tau_{max} = \frac{Kr}{2}\,\sigma_c \frac{E_F}{\bar{E}}\,\tanh K\frac{\ell}{2} = \frac{Kr}{2}\,\sigma_{Fmax} \qquad (29)$$

As shown in Figure 12.12, the plastic region is extended from both ends of the fiber with an increase in the interfacial shear stress. If the interfacial shear stress in the plastic region is given as τ_0, the interfacial shear stress in the residual elastic region ($0 \leq z \leq g$) is given as zero, and the difference between the strains of the fiber and the matrix is neglected, the stresses in the elastic and plastic regions are shown, respectively, as follows:

$$\bar{\sigma}_F = \frac{\sigma_c E_F}{\bar{E}} \qquad 0 \leq z \leq g$$

$$\bar{\sigma}_F = 2\tau_0\,\frac{\ell - z}{r} \qquad g \leq z \leq \frac{\ell}{2} \qquad (30)$$

From the conditions of continuity of stress at z = g, the range of the plastic region is given by

$$\frac{\ell}{2} - g = \frac{r}{2} \frac{E_F}{E} \frac{\sigma_c}{\tau_0} \tag{31}$$

When interference between the fibers is considered, the analysis of the plastic region is extremely complex. As shown in Figure 12.12, when interference exists between the fibers, the boundary between the elastic region and the plastic region is linear. That is, the plastic region first develops at the ends of the fiber and is then extended to reach the adjacent fiber. Thereafter, the plastic region is extended along the fiber and comes to occupy the entire matrix. Although elastoplastic stress analysis is conducted as described above, interfacial stress is also studied by analytical techniques in many cases.

Here an example is shown in which the analysis is carried out using the amplified model of the fiber and the matrix shown in Figure 12.13 [14]. In this case, the assumptions used for the analysis are as follows:

1. The adhesion at the interface of the fiber and the matrix is complete.
2. The vertical stress σ_m of the matrix in the direction of the fiber axis is much lower than the vertical stress σ_f of the fiber.
3. The load is transmitted to the matrix through interfacial shear stress.
4. The fiber is not buckled.
5. Interference with the stress field between the fibers is extremely small.

By solving the balanced equation of the stress components, based on these assumptions, the following equation is obtained:

$$\sigma_f = \sigma \cos^2 \alpha \frac{E_f}{E_a} \left(1 - \frac{\cosh \eta_z}{\cosh \eta \ell_f} \right) \tag{32}$$

Further, the interfacial shear stress is given by

$$\tau = Q \sinh \eta_z + C \tag{33}$$

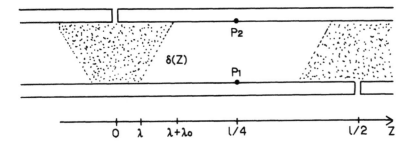

Figure 12.12 Plastic regions growing at fiber ends. (Ref. 12.)

where

$$Q = \frac{G_m \varepsilon^2}{\eta^3 (r_m - r_f) \cosh \eta \ell_f} \left[\frac{r_f^2 \cos \alpha}{E_a (r_a^2 - r_f^2 \cos \alpha)} + \frac{1}{E_f} \right]$$

$$\eta^2 = \frac{2G_m}{(r_m - r_f) r_f} \left[\frac{r_f^2 \cos \alpha}{E_a (r_a^2 - r_f^2 \cos \alpha)} + \frac{1}{E_f} \right]$$

$$\varepsilon^2 = \frac{2\sigma G_m r_a^2 \cos^2 \alpha}{E_a r_f (r_m - r_f)(r_a^2 - r_f^2 \cos \alpha)}$$

In particular, when $r_f \ll r_a$,

$$\eta^2 = \frac{2G_m}{E_f (r_m - r_f) r_f} \qquad \varepsilon^2 = \frac{2\sigma G_M \cos^2 \alpha}{E_a r_f (r_m - r_f)}$$

$$Q = \frac{G_m \varepsilon^2}{E_f \eta^3 (r_m - r_f) \cosh \eta \ell_f}$$

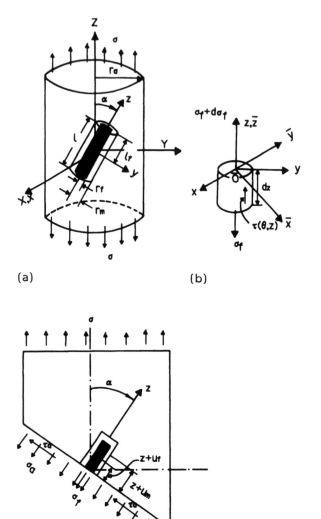

(a)

(b)

(c)

Figure 12.13 Representative volume element. (Ref. 14.)

E_f is the longitudinal elastic modulus of the fiber, G_m the transverse elastic modulus, and E_a the average longitudinal elastic modulus of the model.

The boundary condition for the interfacial shear stress is

$$\tau = -\sigma \sin \alpha \cos \alpha \sin \theta \tag{34}$$

Hence, by using this, at $z = 0$, equation (33) is written as

$$\tau = Q \sinh \eta_z - \sigma \sin \alpha \cos \alpha \sin \theta \tag{35}$$

Diameter d_r of the fiber, aspect ratio a_f of the fiber, content by volume V_f of the fiber, and ratio R of the length of the fiber to that of the matrix are defined, respectively, as

$$a_f = \frac{\ell_f}{d_f} \qquad V_f = \frac{r_f^2 \ell_f}{r_m^2 \ell} \qquad R = \frac{\ell - \ell_f}{\ell_f}$$

In this case, the following equations are given:

$$\eta \ell_f = a_f \left[\frac{2G_m / E_f}{[1/\sqrt{V_f(1 + R)} - 1]} \right]^{1/2}$$

$$Q = \frac{\sigma(\cos^2 \alpha) G_m}{E_a [(2G_m / E_f)(1/\sqrt{V_f(1 + R)}) - 1]^{1/2} \cosh \eta \ell_f}$$

For SMC material consisting of 30% polyester, 30% chopped fibers, and 40% calcium carbonate, the results of calculations on fiber stress and interfacial shear stress are as shown in Figures 12.14 to 12.16.

Figure 12.14 shows $\sigma_{f,max}(z = 0)$ and $\tau_{max}(|z| = \ell_f)$ as a function of the aspect ratio a_f of the fiber. It can be seen that a_f has a very limited effect on τ_{max}, but a large effect on σ_{max}, a_f ranging from 0 to 50. Figure 12.15 shows a relationship between the fiber content of SMC, and σ_{max} and τ_{max}. Both σ_{max} and τ_{max} decrease with increase in fiber content. This indicates that the more fibers play a part in the transmission of the load. Figure 12.16 shows the relationship between τ and α when z and θ are varied. When $\theta = \pm 90°$, τ has its maximum value at $30° \leq \alpha \leq 60°$, and when $\theta = 0°$, τ becomes maximum at $\alpha = 0°$.

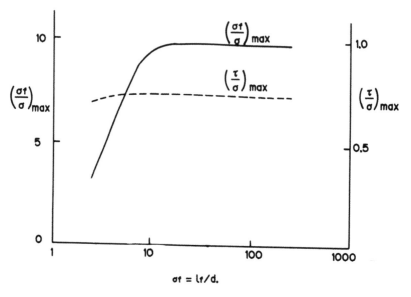

Figure 12.14   Plot of $\sigma_{f,max}$ and τ_{max} as a function of α_f. (Ref. 14.)

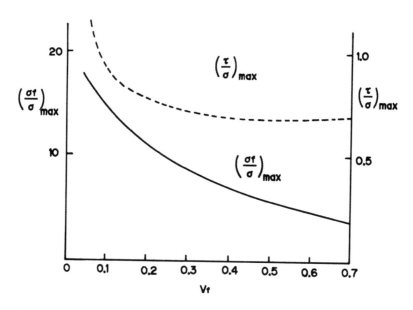

Figure 12.15   Plot of $\sigma_{f,max}$ and τ_{max} as a function of V_f. (Ref. 14.)

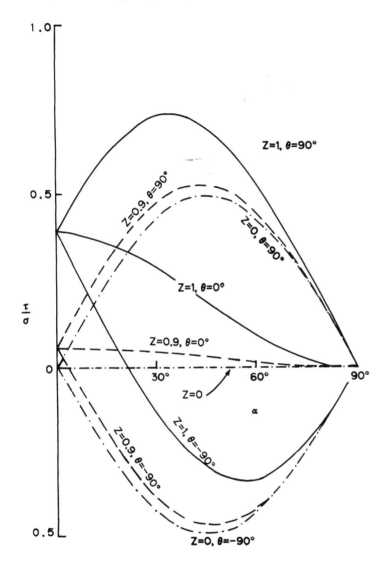

Figure 12.16 Plot of τ/σ for SMC-30 as a function of α. (Ref. 14.)

12.3   MEASUREMENTS OF INTERFACIAL STRENGTH

When the interfacial strength of fiber-reinforced composite material
is in question, factors considered include the interfacial strength
between the fiber and the matrix and the interfacial strength
between the laminates, and if the cloth materials or unidirectional
reinforcing materials are laminated.   Such interfacial strength relates to the interfacial binding force between the fiber (reinforcing
material) and the matrix.   The factors concerned with this strength
are thus closely related not only to the mechanical properties of the
matrix and the fiber (reinforcing material), but also to the physico-
chemical properties.   We will describe a few experimental methods
employed to measure the composite material and interfacial strength
thus obtained.

12.3.1   Interfacial Strength Between
Single Fiber and Matrix

The matrix is charged in the button-form instrument shown in
Figure 12.17, and the fiber is buried in this matrix [15].   After
the matrix is cured, the fiber is pulled out.   The interfacial
strength of the fiber and the matrix is measured by the force required to pull the fiber out.   If fiber of more than a certain
length is buried in the matrix, the fiber breaks rather than pulling
out of the matrix.
    When the maximum buried length at which the fiber pulls out
is L_c, the interfacial shear strength between the fiber and the
matrix is τ, the tensile strength of the fiber is σ, and the diameter
of the fiber is d_f, the following equations are derived from the
balance of the shear strength of the portion where the fiber is
buried:

$$\frac{1}{4}\,\pi d_f^2 \sigma = \pi d_f L_c \tau$$

$$\frac{L_c}{d_f} = \frac{1}{4}\,\frac{\sigma}{\tau}$$

(36)

These equations are derived under the assumption that the interfacial shear stress is uniform.
    Figure 12.18 shows the result of a pull-out test carried out us-
ing a model consisting of a copper matrix with a tungsten wire
buried therein [16].   The ordinate indicates the stress applied to
the tungsten wire and the abscissa indicates the ratio of the buried
length ℓ of the tungsten wire to the wire diameter d.

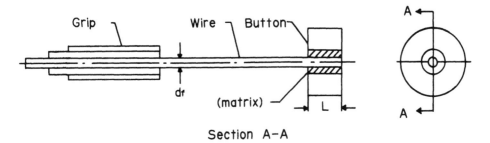

Section A-A

Figure 12.17   Schematic representation of the button-form instrument.   (Ref. 15.)

Cases in which the end face of the tungsten wire is both adhered and not adhered are shown in the graph.   These cases show that a greater force is required to pull out tungsten wire whose end face is adhered.   When the buried length of the tungsten wire is several times as long as the wire diameter, the wire breaks without pulling out.

Figure 12.19 illustrates a similar method used to measure the interfacial strength between the fiber and the matrix [17].   In this case, the fiber is buried in a disk composed of the matrix, and a load is applied to the disk.   The interfacial shear strength is determined by the relationship between the amount of transfer between the disk and the load.   In method (a), a load is applied to the disk in the longitudinally direction of the fiber, and the shear stress is allowed to act on the interface in that direction.   In method (b), the disk is fixed and a torsional load is applied to the fiber that causes the shear stress to be applied in the peripheral direction of the fiber.

As shown in Figure 12.20, the maximum value of the load is considered to be the interfacial strength, taking into account the relation between the load applied to the disk and the transfer amount of the crosshead.   The load that shifts the disk along the fiber beyond the maximum load corresponds to the frictional force between the fiber and the matrix.   This frictional force results from the residual stress in the interface.

Similar results have been obtained by similar experiments in which Kevlar fiber is buried in epoxy resin [18,19].   Table 12.1 shows the interfacial shear strengths obtained by methods (a) and (b) wherein a polyester resin is used as the matrix and a glass rod is used as the fiber.   This result indicates that the value obtained by method (b) is a little higher than that obtained by method (a).

Figure 12.21 illustrates a method for applying a compressive load to the end faces of blocks and of various forms composed of

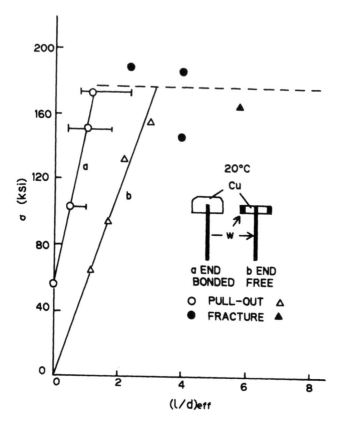

Figure 12.18 Fiber stress at pull-out (open points) or fiber fracture (filled points) as a function of twice the embedded length
divided by the diameter at room temperature. Specimens with fiber
ends bonded followed (a) and unbonded (b). (Ref. 16.)

resin in which a single fiber is buried [17,20]. The test piece
shown in part (a) is a rectangular prism matrix in which fiber is
buried, and a compressive load is applied to it in the direction of
the fiber axis. The difference between the fiber and the matrix
causes shear stress on the end face of the fiber. The interfacial
shear strength is measured by the compressive load value at which
the fiber peels off from the matrix when the load is increased.
Between the compressive stress σ_c applied to the test piece and
the interfacial shear strength between the fiber and the matrix,
the following relationship holds [21]:

$$\tau = 2.5\sigma_c \tag{37}$$

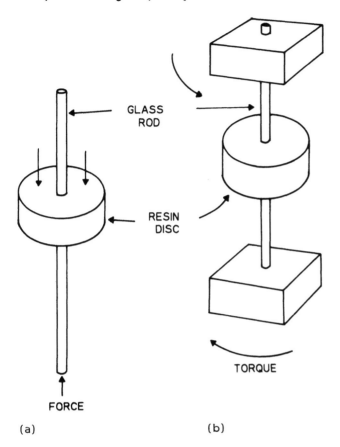

GLASS
ROD

RESIN
DISC

FORCE

TORQUE

(a)   (b)

Figure 12.19   Rod and disk specimens.   (Ref. 17.)

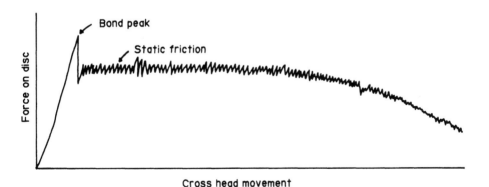

Figure 12.20   Typical load-displacement curve for rod-disk joint
strength specimen.   (Ref. 17.)

Table 12.1   Bond Strength Results for Rod-Disk Specimens[a]

Glass rod diameter (mm)	Surface treatment	Type of test	Bond strength (psi)	Friction load Bond load (%)
2	Acetone cleaned	Push test	614	89
4	Acetone cleaned	Push test	605	94
	Vinyltrichlorosilane	Push test	680	94
	Acetone cleaned	Torsion test	963	

[a]Polymer was Paraplex P43.
Source: Ref. 17.

Similarly, in the model of Figure 12.21(b), when a compressive load is applied to the end face of the test piece, the difference in Poisson's ratio values between the matrix and the fiber causes tensile stress in the interface between them. Interfacial tensile strength can be obtained by the load at which the fiber peels off from the matrix when the compressive load is increased. This interfacial tensile strength σ_a is given as

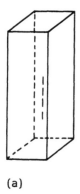
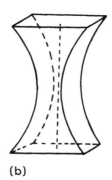
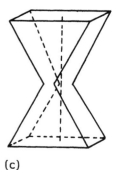

(a)   (b)   (c)

Figure 12.21   Single-fiber specimens: (a) rectangular; (b) curved neck; (c) trapezoidal. (Ref. 20.)

$$\sigma_a = \frac{\sigma_s (\nu_m - \nu_f)E_f}{(1 + \nu_m)E_f + (1 - \nu_f - 2\nu_f^2)E_m} \tag{38}$$

where σ_s is the stress at the minimum cross section of the test piece, ν is Poisson's ratio, E the vertical elastic modulus, and the suffixes f and m represent the fiber and the matrix, respectively.

Similarly, in the model of Figure 12.21(c), when the compressive load is applied to the end face of the test piece, the interfacial shear stress τ_c induced at the minimum cross section between the fiber and the resin is determined by the following equation [22]:

$$\tau_c = - \frac{P}{nX_0^2 T} \frac{E_f}{E_r} \frac{d_f}{4} \tag{39}$$

where P is the compressive load, n the gradient of the test piece, E_f and E_r the elastic module of the fiber and the resin, respectively, and d_f the diameter of the fiber. Table 12.2 shows the interfacial strengths measured by the methods described above.

Figure 12.22 illustrates a method for measuring the interfacial strength between the fiber and the matrix by the use of three fibers [23]. In this case, the fiber to be tested is adhered to two supporting fibers, with the matrix at right angles to the fibers. The interfacial shear strength is determined by the load at which the fiber to be tested is pulled out when a tensile load is applied to the fiber.

As shown in Figure 12.23, if the dimensions of the fiber to be tested and the supporting fibers are determined, the average inter-facial shear strength t_m between the fiber and the matrix can be calculated by the use of the maximum load F_{max} for pulling out the fiber:

$$\tau_m = \frac{F_{max}}{\pi dD} \tag{40}$$

Figure 12.24 shows examples of interfacial shear strengths between the various matrices and glass fiber measured by this method.

Recently, Grande et al. reported an experimental technique and associated analysis to determine the in situ fibre-matrix bond strength in typical as-processed polymer, ceramic, and metal matrix composites using the microdebonding test apparatus [24]. Results are correlated with composite longitudinal and interlaminar shear behavior for carbon and Nicalon fiber-reinforced glasses and glass-ceramics, including the effects of matrix modifications, processing conditions, and high-temperature oxidation embrittlement. The data indicate that significant bonding to improve off-axis and shear properties can be tolerated before the longitudinal behavior becomes brittle.

Table 12.2   Typical Values of Polymer-Glass Joint Strength

Test method	Material type	Glass treatment	Bond strength (psi)
Rod-disk (push test)	Polyester (Paraplex P 43)	Acetone cleaned	605
	Polyester (Paraplex P 43)	Vinyltrichlorosilane	680
Trapezoidal fiber	Polyester (Paraplex P 43) and E glass	Acetone cleaned	1000
	Epoxy (Epon 828)	Acetone cleaned	3000–3500
Curved neck fiber	Polyester (Selectron 5026) and E glass	Heat cleaned	750
	Polyester (Selectron 5026) and E glass	2% A172 in polymer	1220
	Epoxy (Epon 828)	Toluene cleaned	>1540

Source: Ref. 17.

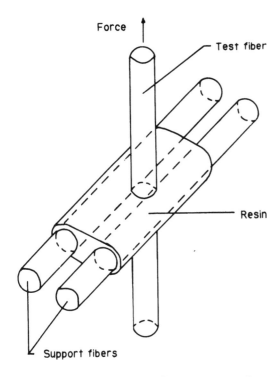

Figure 12.22   Schematic representation of the three-fiber method, used to determine fiber/resin bond strength.   (Ref. 23.)

12.3.2   Measurements of Interfacial Strength Between Fibers or Laminates

The interfacial strength is measured by a method wherein fracture takes place by shear or tensile load along the direction of fiber or the laminate [25].   That is, the interlaminar shear strength is measured by a three-point bending test of the short beam, as shown in Figure 12.25.   For the beam to be tested, a unidirectionally fiber-reinforced material or laminated material is used.   In this case the shear stress becomes maximum at the center face x-x' of the beam, and its value τ_{max} is calculated by use of the applied load P:

$$\tau_{max} = \frac{3P}{4ab} \qquad (41)$$

where a is the thickness of the test piece and b is its width.

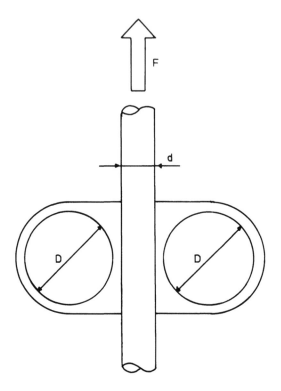

Figure 12.23 Definition of values used to measure the shear
strength of a fiber-resin boundary surface in the three-fiber
method. This method does not account for the wetting between
fiber and resin. (Ref. 23.)

The maximum value σ_{max} of the tensile stress induced in the beam
in the direction of the fiber axis appears at point O in Figure 12.25,
and its value is given as

$$\sigma_{max} = \frac{3PS}{2a^2 b} \tag{42}$$

Whether the tensile fracture takes place at point O, or shear fracture
occurs at the center face x-x', depends on the thickness a of the test
piece and the distance S between the supporting points. When the
shear strength is measured, the test piece must be selected so that
the fracture takes place at the center face x-x'. The dimensions of
the test piece are specified by ASTM as shown in Table 12.3.
    As described in Chapters 3 through 7 with respect to composite
materials, the interaction between the fiber and the matrix is affected considerably by surface treatment or modification of the fiber

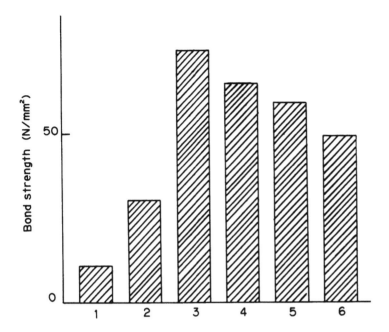

Figure 12.24 Bond strengths of commercial resins in the three-fiber test with A-glass fiber. 1, Loctite Multibond; 2, Scotch 3M glass adhesive; 3, Loctite glass adhesive (UV-cured); 4, Tammer F3 (phenol-formaldehyde resin); 5, Plastic Padding Super Adhesive; 6, Super Epoxy. (Ref. 23.)

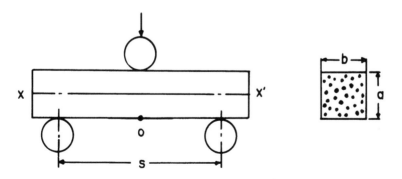

Figure 12.25 Short-beam shear test. (Ref. 25.)

Table 12.3   Dimension of Specimens for the Short-Beam Test

	Span/thickness	Length/thickness
Woven cloth reinforcement	5	7
Continuous glass filaments	5	7
Silica fibers (continuous)	4	6
Graphite yarn	4	6
Carbon yarn	5	7
Boron filaments	4	6
Steel wire	5	7

Source: Ref. 25.

Table 12.4   Effect of Fiber Surface Treatment on Short-Beam Shear Strength

Resin	Filabond 8000
Curing conditions	Gelatin at room temperature, postcured at 120°C for 5 h
Depth/span ratio	1:5.5

Surface treatment	Short-beam shear strength (MNm^{-2})
A Water only	26.8
B 0.3% Silane A174 in water, pH adjusted to 3.5–4.0 with acetic acid	65.9
C 0.3% Silane A153 in water, pH 3.5–4.0	23.6
D 0.3% Morpan TPB in water	25.5
E 0.3% Silane A174 and 0.1% Morpan TPB in water, pH 3.5–4.0	60.7
F 0.3% Silane A174. 0.1% Morpan TPB and 5% poly(vinyl acetate)	56.7

Source: Ref. 26.

or matrix. Table 12.4 shows the interlaminar shear strengths of unidirectional reinforcing materials consisting of polyester resin and glass fibers variously surface-treated [26]. It can be seen that the interlaminar shear strength is influenced significantly by the treatment agent.

As shown in Figure 12.26, two grooves are cut on the laminated plate and a tensile load is applied, which the interfacial shear stress is measured [27]. In this case, the mean value τ of the shear stress is determined by the equation

$$\tau = \frac{P}{W\ell} \tag{43}$$

where P is the load, W the width of the test piece, and ℓ the distance between the grooves. Table 12.5 shows a relationship between the depth of the groove and the shear stress for the laminated plate of aramid fibers (65 vol %) and epoxy resin.

Other methods have also been proposed, as shown in Figure 12.27. In Figure 12.27(a) a cylinder reinforced with fiber in the peripheral direction is used as the test piece. If the tensile or compressive load applied to the test piece is P and the torsional moment is T, the shear stress τ induced in the test piece is given as

$$\tau = \frac{T}{2\pi r_m^2 t} \tag{44}$$

where t is the plate thickness of the cylindrical test piece and r_m is the mean of the outer diameter and the inner diameter of the cylinder.

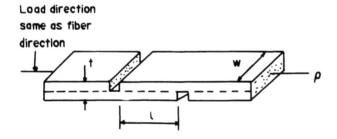

**Load direction
same as fiber
direction**

Unidirectional grooved laminate, interlaminar tensile shear

Figure 12.26   Schematic representation of specimen for shear test. (Ref. 27.)

Table 12.5 Effect of Groove Depth on the Maximum Interlaminar Shear Stress Calculated from Tensile Shear of Grooved Laminates

Specimen	Mean shear stress between grooves (MPa)	Number of specimens	Maximum shear stress (MPa)
With grooves cut to center ply or slightly under	15.2	5	14.03
	16.7	5	23.05
	14.0	4	28.42 $A_v = 23.46$
	10.9	5	23.31
	10.7	5	28.50
With grooves cut through center ply or slightly over	14.14	3	12.26
	11.73	4	15.15
	8.49	4	14.70 $A_v = 15.52$
	8.18	4	17.51
	7.44	3	17.97
With overcut grooves	5.20	5	4.82
	5.42	5	6.86
	5.05	5	8.71 $A_v = 7.71$
	4.12	5	9.00
	3.76	5	9.17

Source: Ref. 27.

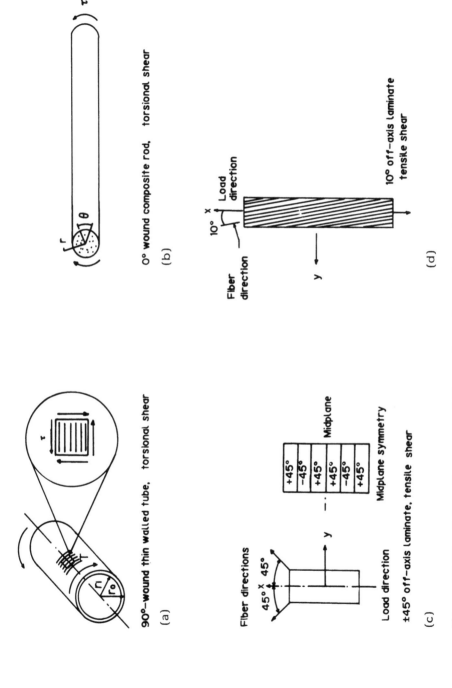

Figure 12.27 Schematic representations of specimens for various shear tests. (Ref. 27.)

In Figure 12.27(b) when a torsional moment (T) is applied to a columnar test piece in which the reinforcing fiber is arranged in the axial direction, a maximum shear stress (τ_m) is induced on the surface of the column as given by the equation

$$\tau_m = \frac{2T}{\pi r^3} \tag{45}$$

where r is the radius of the column.

Figure 12.27(c) shows a test piece in which the reinforcing fibers are laminated at angles of ±45° to each other. When the average tensile stress is σ_x, the shear stress (τ) is given as

$$\tau = \frac{\sigma_x}{2} \tag{46}$$

Figure 12.27(d) shows a test piece in which the angle between the tensile direction and the fiber axis is 10°. The following relationship exists between the tensile stress σ_x and the shear stress:

$$\tau = 0.171\sigma_x \tag{47}$$

Table 12.6 shows the shear characteristics determined for the test pieces described above.

Figure 12.28 illustrates a method for measuring the interfacial tensile strength of a laminated plate, wherein a metal instrument is adhered in the laminated plate and a tensile load applied [28]. A high stress concentration occurs at the end faces of the laminated plate, its degree depending on the properties of the single plates to be laminated and the order in which they are laminated. For this reason, in the test piece of the model shown in Figure 12.28(b), the end faces of the instrument are cut off and those parts filled with the resin.

The strength of a laminated interface measured by use of a ring-form laminated test piece is shown in Figure 12.29 [29]. When the tensile or compressive load, torsion, or a combination of these is loaded on the test piece through this instrument, the vertical stress (i.e., the tensile or compressive stress), the shear stress, or a combination can be loaded on the interface of the laminate. From the load at which the test piece breaks, the interfacial strength can be obtained using equation (44). Figure 12.30 shows an example of a laminated material consisting of glass cloth and epoxy-acrylate resin.

With respect to fiber-reinforced materials, the interfacial strength between a single fiber and a matrix, and the strength of

Table 12.6   Comparison of Shear Properties Obtained from Various Test Methods[a,b,c]

	Shear failure stress (MPa)	Shear strain at failure stress (%)	Shear modulus at 0.5% shear strain (MPa)
Torsional shear of 90°-wound thin tube	31.3 ± 2.0	2.02 ± 0.20	1744 ± 39
Torsional shear of a 0°-wound composite rod			
Strain measured with gauges	31.3 ± 2.8	1.82 ± 0.20	1965 ± 89
Strain from angle of twist	32.5 ± 3.0	2.06 ± 0.17	1758 ± 21
Tensile shear of ±45° off-axis laminates			
Midplane symmetry (seven-layer)	29.4 ± 0.7	1.73 ± 0.05	1923 ± 113
Midplane symmetry (seven layers repeated)	27.9 ± 1.6	1.68 ± 0.05	1889 ± 20
Nonsymmetrical (eight-layer)	31.7 ± 1.4	1.89 ± 0.06	1875 ± 90
Tensile shear of 10° off-axis laminates			
Coated edges	19.4 ± 1.1	0.97 ± 0.08	2082 ± 188
Bare edges	19.1 ± 2.9	1.03 ± 0.21	1903 ± 191

[a]Material: 65 vol % aramid fiber in epoxy No. 1.
[b]Plus or minus values indicate 95% confidence limits.
[c]See Table 12.5 for details.
Source: Ref. 27.

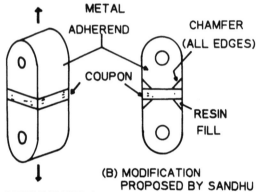

Figure 12.28   Flatwise tension specimen.   (Ref. 28.)

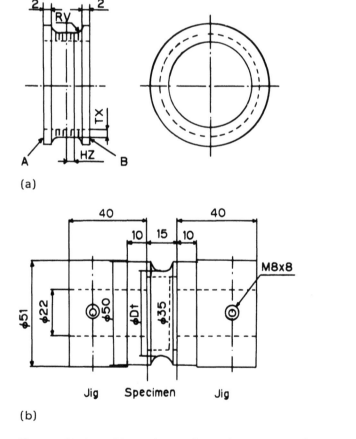

(a)

(b)

Figure 12.29   Dimensions of specimens.   (Ref. 29.)

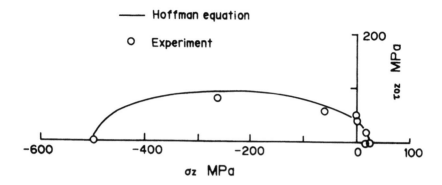

Figure 12.30   Interlaminar strength under combined stress (axial tensile or compressive stress/axial twist stress = 2:1). σ_z, axial stress (stress in + direction is tensile; stress in − direction is compressive); τ_{oz}, shear stress. (Ref. 29.)

the laminated interface of a laminated material, have been described from a phenomenalistic viewpoint. In many cases, the interfacial strength of fiber-reinforced material is generally lower than that of the matrix or reinforcing fiber. Accordingly, when fiber-reinforced material is used as the structure, the structural strength sometimes depends on the properties of the interface, with its lower strength, rather than on those of the fiber, with its higher strength. In such cases, the effect produced by using a reinforcing fiber is insufficient. To allow the fiber-reinforced material literally to play a part as reinforcing material, improving the interfacial strength seems to be the most important consideration.

REFERENCES

1. K. Kubo, M. Koishi, and T. Tsunoda, Eds., *Composite Materials and Interface*, Sōgō Gijutsu Shuppan, Tokyo, 1986, pp. 37−79.

2. T. Fujii, A Trend in Research on the Strength of Reinforced Plastics, *J. Soc. Mater. Sci. Jpn.*, *19*, 937 (1970).

3a. T. Ishikawa and S. Kobayashi, Stress Analysis of Unidirectional Fiber-Reinforced Composites, *J. Jpn. Soc. Compos. Mater.*, *2*, 126 (1976).

3b. T. Ishikawa, Strength of Unidirectional Fiber-Reinforced Composite Materials, *J. Jpn. Soc. Aerosp.*, *26*, 44 (1978).

4. A. Uemura, H. Kawai, H. Maki, and N. Watanabe, eds., *New Composite Materials and Advanced Technology*, Tokyo Kagaku Dozin Co. Ltd., Tokyo, 1986, p. 171.

5a. T. Fujii and M. Zako, *Break and Dynamics of Composite Materials*, Jitsukyo Shuppan, Tokyo, 1985, pp. 96–102.

5b. H. L. Cox, F. R. Ae, and A. M. I. Mech, The Elasticity and Strength of Paper and Other Fibrous Materials, *Br. J. Appl. Phys.*, *3*, 72 (1952).

6. J. O. Outwater, Jr., The Mechanics of Plastics Reinforcement in Tension, *Mod. Plast.*, *33*, 156 (1956).

7. N. F. Dow and B. W. Rosen, Evaluations of Filament-Reinforced Composites for Aerospace Structural Applications, NASA CR-207, National Aeronautics and Space Administration, Washington, D.C., 1965.

8. B. W. Rosen, Mechanics of Composite Strengthening, *Fibre Compos. Mater.*, *ASM72*, 75 (1965).

9. W. R. Tyson and G. J. Davies, A Photoelastic Study of the Shear Stresses Associated with the Transfer of Stress During Fibre Reinforcement, *Br. J. Appl. Phys.*, *16*, 199 (1965).

10. M. R. Riggot, A Theory of Fiber Strengthening, *Acta Met.*, *14*, 1429 (1966).

11. A. Kelly and W. R. Tyson, *High Strength Materials*, John Wiley & Sons, Inc., New York, 1965, p. 578.

12. B. Schultrich, W. Pomepe, and H. J. Weiss, The Influence of Fiber Discontinuities on the Stress-Strain Behaviour of Composites, *Fibre Sci. Technol.*, *11*, 1 (1978).

13. B. Lauke and B. Schultrich, Deformation Behavior of Short-Fiber Reinforced Materials with Debonding Interfaces, *Fibre Sci. Technol.*, *19*, 111 (1983).

14. C. T. Chon and C. T. Sun, Stress Distributions Along a Short Fiber in Fiber Reinforced Plastics, *J. Mater. Sci.*, *15*, 931 (1980).

15. C. C. Chamis, *Composite Materials*, Vol. 6, Academic Press, Inc., New York, 1974, p. 75.

16. A. Kelly and W. R. Tyson, Tensile Properties of Fiber-Reinforced Metals, Copper/Tungsten and Copper/Molybdenum, *J. Mech. Phys. Solids*, *13*, 329 (1965).

17. L. J. Broutman, Glass-Resin Joint Strengths and Their Effect on Failure Mechanisms in Reinforced Plastics, *Polym. Eng. Sci.* (July), 263 (1966).

18. L. S. Penn, F. A. Bystry, and H. J. Marchionni, Relation of Interfacial Adhesion in Kevlar/Epoxy Systems to Surface Characterization and Composite Performance, *Polym. Compos.*, *4*, 26 (1983).

19. L. S. Penn and S. M. Lee, Interpretation of the Force Trace for Kevlar/Epoxy Single Filament Pull-Out Tests, *Fibre Sci. Technol.*, *17*, 91 (1982).

20. J. B. Shortall and H. W. C. Yip, The Interfacial Bond Strength in Glass Fiber-Polyester Resin Composite Systems, *J. Adhes.*, *7*, 311 (1976).

21. L. T. Broutman, Interfaces in Composites, *ASTM STP-452*, American Society for Testing and Materials, Philadelphia, 1969, p. 29.

22. R. D. Mooney and F. J. McGarry, *14th Annu. Tech. Manage. Conf.*, Reinforced Plastic. Div. SPI, *12-E*, 1 (1959).

23. P. Järvelä, K. W. Laitinen, J. Purola, and P. Törmälä, The Three-Fiber Method for Measuring Glassfiber to Resin Bond Strength, *Int. J. Adhes. Adhes.*, *3*, 141 (1983).

24. D. H. Grande, J. F. Mandell, and K. C. C. Hong, Fiber-Matrix Bond Strength Studies of Glass, Ceramic, and Metal Matrix Composites, *J. Mater. Sci.*, *23*, 311 (1988).

25. American Society for Testing and Materials, Standard Test Method for Apparent Interlaminar Shear Strength of Parallel Fiber Composites by Short Beam Method, *ASTM D-2344*, ASTM, Philadelphia, 1976.

26. H. W. C. Yip and J. B. Shortall, The Interfacial Bond Strength in Glassfiber-Polyester Resin Compsoite Systems, *J. Adhes.*, *8*, 155 (1976).

27. C. C. Chiao, R. L. Moore, and T. T. Chiao, Measurement of Shear Properties of Fiber Composites, *Composites*, *8*, 161 (1977).

28. A. Harris and O. Orringer, Investigation of Angle-Ply Delamination Specimen for Interlaminar Strength Test, *J. Compos. Mater.*, *12*, 285 (1978).

29. M. Funabashi and K. Ikegami, Interlaminar Strength Test of Glass Cloth Laminated Plastics Under Combined Stress, *J. Soc. Mater. Sci. Jpn.*, *33*, 1566 (1984).

Index

About the Authors

RYUTOKU YOSOMIYA is Professor in the Department of Industrial Chemistry, Chiba Institute of Technology, Chiba, Japan. Dr. Yosomiya is the author or coauthor of three books and 125 articles and book chapters. A member of the Society of Polymer Science, Japan, and Ceramic Society, Japan, Dr. Yosomiya received the Doctor of Engineering degree (1965) from Kyoto University, Japan.

KIYOTAKE MORIMOTO is Research Manager of the Department of Composites, Tokyo Research Center, Nisshinbo Industries, Inc., Tokyo, Japan. The author or coauthor of some 20 papers focusing on polyurethane composites, Dr. Morimoto is a member of the American Chemical Society; Society of Polymer Science, Japan; Society of Materials Science, Japan; and American Society for Metals, International. He received the Doctor of Engineering degree (1984) from Kyoto University, Japan.

AKIO NAKAJIMA is Professor in the Department of Applied Chemistry, Osaka Institute of Technology, Osaka, Japan. He is also Professor Emeritus of Polymer Chemistry, Kyoto University, Kyoto, Japan. Dr. Nakajima has published six books and over 300 articles and book chapters. He is a member of the American Chemical Society, New York Academy of Sciences, Macromolecular Division of IUPAC (national representative), Society of Polymer Science, Japan (president, 1978–80), and Japanese Society for Biomaterials (president, 1984–88). Dr. Nakajima received the Doctor of Engineering degree (1951) from Kyoto University, Japan.

YOSHITO IKADA is Director and Professor of the Research Center for Medical Polymers and Biomaterials, Kyoto University, Japan. Dr. Ikada focuses his work on polymers, polymer surfaces, and biomaterials, and he has published three books in these areas. Among the professional organizations he belongs to the American Chemical Society, Society of Polymer Science, Japan; and Japanese Society for Biomaterials. He received the Doctor of Engineering (1963) and Doctor of Medical Science (1984) degrees from Kyoto University, Japan.

TOSHIO SUZUKI is Director and General Manager of the R&D Division, Nisshinbo Industries, Inc., Tokyo, Japan. Mr. Suzuki holds 40 patents in the areas of thermoset resins, polyurethanes, and plastic films. He is a member of the Society of Polymer Science, Japan, and Society of Fiber Science and Technology, Japan. Mr. Suzuki received the B.S. degree (1957) in chemistry from Tokyo Institute of Technology, Japan.